A NETWORK ORANGE

A NETWORK ORANGE

LOGIC AND RESPONSIBILITY
IN THE
COMPUTER AGE

Richard Crandall
Marvin Levich

With a Foreword by Howard Rheingold

COPERNICUS
AN IMPRINT OF SPRINGER-VERLAG

Published in the United States by Copernicus,
an imprint of Springer-Verlag New York, Inc.

Copernicus
Springer-Verlag New York, Inc.
175 Fifth Avenue
New York, NY 10010

Library of Congress Cataloging-in-Publication Data

Crandall, Richard E., 1947–
    A network orange : logic and responsibility in the Computer Age /
Richard Crandall, Marvin Levich.
        p.   cm.
    Includes bibliographical references and index.
    ISBN 0-387-94647-0 (alk. paper)
    1. Computers and civilization.   2. Computers—Moral and ethical
aspects.   I. Levich, Marvin.   II. Title.
QA76.9.C66C73   1997
303.48′34—dc21   97-33271

Acquiring editor: Allan M. Wylde
Manufactured in the United States of America.
Printed on acid-free paper.

9   8   7   6   5   4   3   2   1

ISBN 0-387-94647-0   SPIN 10526016

ABOUT THE TITLE

The title of this book is derived from the Cockney folk expression "as queer as a clockwork orange," which means, in the words of author Anthony Burgess, "the queerest thing imaginable." We take the folk expression to describe an entity that, although it appears ordered and natural on first inspection, is ultimately chaotic within. A secondary, anthropomorphic meaning has thrived since the appearance of Burgess's horrific novel *A Clockwork Orange,* in which the human mechanism, organic but capable of brutal, robotic behavior, is the clockwork orange. We intend the original folk meaning: The modern network, already vast but integral to an even vaster computer revolution, is a strange and chaotic thing, at the fringe of the unimaginable, giving rise to profound problems of logic and responsibility.

CONTENTS

FOREWORD

Computer technology has become a mirror of what we are and a screen on which we project both our hopes and our fears for the way the world is changing. Earlier in this century, particularly in the post–World War II era of unprecedented growth and prosperity, the social contract between citizens and scientists/engineers was epitomized by the line Ronald Reagan promoted as spokesman for General Electric: "Progress is our most important product." In more recent decades, post-Chernobyl, post-Challenger, post-Bhopal, post-Microsoft, the social contract has undergone a transformation. More people are uncertain, fearful, and downright opposed to the notion that more technology guarantees a better life. What is a "better life"? Who benefits and who loses when new technologies change the way we live, work, learn, and play? Who has a say in the way technologies are designed and deployed? Where are we going, are we sure we want to go there, and who has the power to do anything about it?

From the early days of the railroads, into the era of electrification, through the McLuhan age, much of the discourse about technology has been hype, utopianism, and what some historians have called "the rhetoric of the technological sublime." We have discovered, however, that not all people benefit economically or politically from technological change. We now know that democratic discourse has suffered from the transformation of the public sphere into a marketplace where issues and candidates are packaged and sold as commodities. And we have learned that many of the claims for new technologies are made by marketeers, not objective observers.

The computer is perhaps the most important projection screen for our hopes and fears about where we are heading. M.I.T. sociologist Sherry

Turkle calls the computer "an object to think with"—to reflect on the kind of people we are becoming and on our deepest feelings about the shadow side of technology.

Artificial intelligence, multimedia, virtual reality, nanotechnology, information superhighways, and educational revolutions have been hyped to the point where nobody is sure where the hyperbole leaves off and technology forecasting begins.

Most important, it has taken decades to realize that the questions engendered by technological change are precisely the questions that classical liberal arts education was designed to raise: What is the good life? How should we treat one another? Does history have a direction? Does human life have a purpose? Is there a metaphysical and logical ground for philosophical and ethical questions, or are they by their nature unanswerable? How do people govern themselves when their opinions about the good life and ethical behavior differ?

These are the questions philosophers have been asking for thousands of years. For a couple of centuries, philosophy and the other humanities were relegated to the realm of interesting antiquities. Science and technology were the privileged and powerful means of determining truth. Now that we have come up against the limits of science and technology as truth-discovering tools, we are returning to the humanities. If you want a discussion of technology that has any depth, you can't trust scientists and engineers alone. You have to invite philosophers to the table.

Literary critics create "works of art based on works of art." Literary critics understand and appreciate literature. Yet most technology criticism, and almost all technology criticism written for lay readers, has been written by those who feel technology is a problem, not a solution. The mass media have contributed to the polarization of this debate. Are computers going to herald an era of democratization and labor-saving devices? Or can they only fold, spindle, and mutilate the human spirit? Debates between those who see technology as white-hatted and those who see it as black-hearted make for dramatic television, but they ignore the important questions that lurk in the gray areas. What are computers good at? What are they not so good at? What is computer technology likely to achieve soon? What is it unlikely to achieve ever? Can we forecast and guide the development of new technologies? Can we understand the impact of new computer technologies before they are fully developed and deployed?

The value of the present volume lies precisely in the fact that one author is a technologist, the other a philosopher. It takes some courage for a philosopher to make a critique of technology and for a computer scientist to team up with a philosopher. Both authors have worked for many years with computer technology in the classroom, as a research instrument, and as a productivity tool. They have also been deeply involved with the actual creation of computer hardware, academic software, and network technology. Together, they have attempted to sort through the hype and create a work of criticism that explores the shades of gray. It's a start. Let us hope that A Network Orange inspires others in the humanities to look at the philosophical and ethical bases of technology debates and that it leads to broader, deeper discourse about the ways computer-related technologies affect our lives.

Howard Rheingold
Electric Minds
hlr@well.com
http://www.minds.com

PREFACE

> I t is perhaps paradoxical that, just when in the deepest sense man has ceased to believe in—let alone to trust—his own autonomy, he has begun to rely on autonomous machines, that is, on machines that operate for long periods of time entirely on the basis of their own internal realities. If his reliance on such machines is to be based on something other than unmitigated despair or blind faith, he must explain to himself what these machines do and even how they do what they do. This requires him to build some conception of their internal "realities."[1]

This collection of essays is intended to expose the "computer revolution" as being simultaneously more and less than is commonly acknowledged. We wrote this collection on the belief that it is time for sharp scrutiny of a revolution that too often eludes proper focus. A primary objective of our work is to lay out the issues of responsibility that face us. The chaos inherent in the modern network, not to mention a host of intellectual problems attendant on the grander revolution itself, gives rise to profound problems of responsibility for computer users. Likewise, educators must take responsibility for the intellectual character of the computer

---

[1] J. Weizenbaum, *Computer Power and Human Reason* (New York: Freeman, 1976), p. 9.

revolution, and engineers must once and for all consider the cultural implications of future designs.

Of course there are many advances in the use of computer technology. Some are important, some trivial, some suspect, and some downright dangerous. It is time to separate the technical wheat from the journalistic chaff. Albert Einstein once remarked that those who desire to understand physics should look at what physicists do, not at what they say they are doing. We believe too much attention has been paid to what computer technologists say they are doing, and not enough to what they *are* doing. We wrote this book to expose and, we hope, help correct that imbalance.

The computer revolution is a whirlwind generated by the combined forces of technology and the media. It blows over artisan, technologist, and politician, at once inspiring unrelieved dread and misguided enthusiasm. We wish to cool the heated rhetoric and examine the subject itself. In so doing, we have not tried to provide an encyclopedic account of everything happening in the world of computers. Nor do we address transient issues involving computers currently in vogue or differences between vendors and designs. We focus instead on some major conceptual and technical issues raised by the vaunted revolution. We believe that nothing is inevitable. The dangers can be avoided, and the benefits realized, if we recognize that the ultimate consequences of the computer revolution will be controlled by people rather than machines. But this will be possible only if people begin to understand the machines. Accordingly, we have tried to avoid recondite technical jargon. The jargon too frequently numbs the lay person into resigned—or, worse, complacent—acceptance.

WHAT *A NETWORK ORANGE* IS NOT

This is not a technical book as such. It is a warning about modern technology. But the book is neither an exercise in nostalgia nor a Neo-Luddite polemic. We authors use computers systematically, for scholarly work, collegial communication, or sheer computation. We are not afraid of robots. For example, in one of the essays, we speculate on what it might take for a machine to be deemed "conscious." We do not believe computers have swept away a humane civilization, though we think that without proper control the effects will be lamentable. From time to time we beat some drums, but we are not on a warpath. We try to indicate what aspects of computing are legitimate and potentially beneficial. Several very important tasks are better performed by computers, and the world of the applied sciences has been positively affected by their use. Rather than—as Norbert Wiener puts it—"fall into a certain vertigo" of negativity and unclear perception, we have endeavored to offer some constructive antidotes to the disease of computer dependence. We shall insist that worshippers of technology need to avoid embracing every entertaining boondoggle that masquerades as "the latest thing." We draw our mission in large measure from Wiener's wise words:

> Render unto man the things which are man's and unto the computer the things which are the computer's. This would seem the intelligent policy to adopt when we employ men and computers together in common undertakings. It is a policy as far removed from that of the gadget worshiper as it is from the man who sees only blasphemy and the degradation of man in the use of any mechanical adjuvants whatever to thoughts. What we now need is an independent study of systems involving both human and mechanical elements. This system should not be prejudiced either by a mechanical or antimechanical bias.[2]

Thus *A Network Orange* seeks balance between extremes. We have tried to expose a certain anti-intellectualism that, ironically, has dominated the technical developments of the last 50 years. Yet, although superb technicians can say irresponsible things about what they are doing, we believe they can remain superb technicians. That is our balance.

ORGANIZATION OF THE ESSAYS

Generally speaking, the first three essays can be taken as a "scientist's perspective," whereas the last three essays offer a "philosopher's perspective." Therefore, the reader will find that the flavor of the presentation changes midway through the collection. The first half of the collection addresses points of interest to a technologist; the second half might appeal more to the educator.

ACKNOWLEDGMENTS

We would like to thank Reed College President S. Koblik for his enlightened view of computer-related scholarship; it was he who founded the

---

[2] N. Wiener, *God & Golem, Inc.* (Cambridge, MA: M.I.T. Press, 1964), p. 73.

Center for Advanced Computation at which we created this book. Likewise, for their academic and administrative support of the Center, we thank Dean of Faculty L. Mantel and Vice President–Treasurer E. McFarlane, respectively. We are grateful to Director of Computing & User Services M. Ringle and Director of Academic Computing G. Schlickeiser for the information they provided about the history of academic computing. We thank fellow scholars S. Arch, S. Danon, M. Hawley, and P. Langston for certain facts and literary references. The advice and encouragement of M. Cloutier, D. Tiffany, A. Cruz and J.A. Westberg were very helpful. L. Powell and T. Day provided critical readings of the manuscript. M. Huddleston provided clutch manuscript processing. We thank numerous editors and reviewers, especially W. Frucht and his editing colleagues, for conferring some measure of sobriety upon us. Finally, we are indebted to our publisher A. Wylde and wish to acknowledge in no uncertain terms his virtually infinite patience.

Richard Crandall and Marvin Levich
Center for Advanced Computation
Reed College, Portland, Oregon

A CONSPIRACY
OF PARTS

W hen you buy a computer from company X and the next year attempt to add extra memory, also from company X, you often discover that they cannot be used together; you'll have to buy a new computer as well. If you are foolish enough to buy two or more different kinds of desktop computers for your staff . . . the expensive software that you buy for one will not run on the other. Rarely will this year's new program run with a six-year-old operating system . . . or a ten-year-old machine. A new program may not even run on last year's model of the same machine. And yet every few years each software package will be "updated" to one with more features and different bugs. You hardly dare pass up the relatively inexpensive update package lest the next generation leaves yours obsolete, incompatible, and unsupported.[1]

DOUBLY FLAWED ANATOMICAL DESIGN

This essay concerns the blizzard of parts, collections of parts, and systems of collections of parts—and the corresponding avalanche of software and software systems—into which a user of modern computers is forced to

---

[1] Landauer 1995, pp. 118–119.

1

march. One aspect of the phenomenon, which might be called "perpetual obsolescence," is echoed in the quote at the opening of this chapter. But there are other aspects: the sheer intricacy of the technology and the recurring attempts to break technological barriers by some inspired combination of speed, parts count, and enhanced software.

As for intricacy, consider a theoretical biologist's lament:

> Serious students of cellular and developmental biology confront what may be the gravest epistemological problems ever faced by scientists. These are a direct consequence of the immense, ordered biochemical complexity of organisms.[2]

Students of technology likewise face a grave epistemological problem, that of assessing and rationalizing the continually expanding dependence of modern machinery on more and more parts and on ballooning software. As we shall see, this phenomenon of dependence goes back to the very beginning of computer technology, when the first machines were built as crude replicas of the human brain as it was then conceived. It was thought that a great deal of memory cells and processing machinery (one might say "brain parts") were needed to achieve true computing power. For better or worse, this basic notion still haunts us today.

Our current technology is thus left with at least two unfortunate, possibly intransigent flaws. First, much of the year-to-year advance in computer design reflects the credo of "more, smaller, faster" parts that bring nothing conceptually refreshing yet often add cost. More unfortunate is a second flaw: In spite of being designed as though they were made up of legitimate biological components, our machines do not yet enjoy the tasking redundancy characteristic of living parts. By tasking redundancy we mean not merely extra parts, but parts that genuinely contribute to some task when other parts fail. An infamous telephone network crash that a decade ago brought down communication in the eastern United States was eventually traced to a bug in a single line of computer code. In June of 1996, the European Space Agency's $7 billion rocket *Ariadne* self-destructed less than a minute into its maiden flight. Again a single software bug caused this debacle. It is because of their abject lack of tasking redundancy that gigantic systems can fail on the basis of a single glitch. By contrast, it is unlikely that the failure of a sin-

---

[2] Kauffman 1972.

gle brain neuron would cause you even to slur a word or slip on the sidewalk.

Having marched for five decades into the world of "anatomical" computer design, we yet find ourselves lacking the greatest advantages of true anatomical structures. Our computers have the hallmarks of complexity but offer little of the benefit. There are two possible future scenarios in which this flaw could be righted. In one scenario, we would be able to fabricate so overwhelmingly many parts as to impose a kind of "brute force" solution. In the second scenario, the entire anatomical motif would be re-evaluated so that the very meaning of anatomical design be effectively revised. Deeper biological analogies—certainly deeper than the mere organization of "organs"—would be brought to bear, bestowing on machinery a new kind of redundancy.

Because we can neither halt the parts blizzard nor enjoy the true biological advantages of ever-more-complex parts assemblies, we choose to think of the parts phenomenon as a conspiracy. By this we mean nothing sinister—certainly nothing like the "planned obsolescence" that folklore often imputes to the automobile industry. We shall concentrate instead on the manner in which we have come to this conspirational impasse. Ultimately, we shall consider means that may eventually dissolve the conspiracy and right at least one of those two basic flaws.

COMPUTER TECHNOLOGY AS A PRODUCT OF WORLD WAR

To the technology historian, few things are as appalling as the general public ignorance of the stark fact that the last two world wars provided the spawning ground—and of course the motivation—for modern electronic design. Both AM and FM radio technology (and some aircraft radio methods as well) were invented by Major Edwin H. Armstrong. Furthermore, the original motive for the invention of sensitive radio receivers was the stringent need for air-medium communication during World War I. Other electronic devices, such as microwave, telephony, and audio appliances (stereo FM was yet another invention of Major Armstrong), are likewise products of world war. In fact, it is difficult to find electronic designs that cannot be traced back to war technology.

By the same token, computing machinery has undeniable military origin. In the early years of World War II, the first computers—the American Mark I and the British COLOSSUS—were built for the express purposes of code breaking and ballistic calculations.[3,4] If you find it sad that the emergence of computing is so closely tied to the Second World War, consider the still sadder fact that even when vacuum tubes (with their vastly greater speed than mechanical relay motions) finally entered the design phase, their first big application was a brace of calculations pertinent to development of the atomic bomb.

These military origins show us one way in which the conspiracy of parts perpetuates itself. The first ballistics calculations were neither conveniently fast nor sufficiently deep. The logical thing to do was add more parts, more relays. This technology was to be abandoned outright, of course, as soon as the much faster vacuum tubes could be implemented. But then there were never enough tubes. The operative principle is that within a given era of computerization, the expansion is overwhelmingly that of quantity, of proliferation; once in a while, there may be an improvement in speed. This "punctuated expansion" is as evident now as it was in the World War II era. It is a central theme of this essay that one reason why we find ourselves living in the conspiracy is that, in era after era since before mid-century, our natural impulse for improvement has been to proliferate parts or, at the very least, enhance the parts themselves.

War-occasioned developments, incidentally, were not limited to machine parts. A great deal of numerical analysis—that abstract but pragmatic branch of computer science—likewise arose from world war. The pervasiveness and longevity of such war developments can be impressive. Just one example is the mathematical career of the eminent mathematician Olga Taussky Todd. As an expert on and self-professed lover of matrix theory,[5] she originated, during World War II, certain techniques of estimation that gave rise to methods of numerical analysis still in use today. (She dealt with matrices relevant to the flutter analysis of aircraft wings for the British government.) Even the very style with which computer scientists

---

[3] Malone 1995. These two original machines were electromechanical relay-based; that is, they did not even use vacuum tubes. Both speed and reliability were abysmal. But one cannot underestimate the military imperative: The machines were built, used, and maintained until the tubes came along.

[4] Some authors cite ENIAC, which we discuss later in this essay, as the first computer. In fact, it was the first large-scale vacuum-tube computer (some 18,000 tubes, performing over 300 operations per second) but ENIAC itself had a calculator-sized precursor (Augarten 1996).

[5] See Varga 1996 for an interesting obituary of the remarkable Olga Taussky Todd, with references to further sources of information about her.

imagine and attack abstract problems can be said to have emerged in good measure from world war.

The postwar era saw the origins of modern computing: Parts counts began their inexorable nonlinear growth, tubes were replaced by transistors, and—most important for the present essay—the anatomical motif in computer design was born.

THE BRILLIANCE OF JOHN VON NEUMANN

It has been argued that modern computing machinery was designed—or at least determined stylistically—by one person, the incomparable John von Neumann. It is possible to trace the anatomical design phenomenon back to his simultaneous command of mathematics, systems (with a view to both game theory and physiological systems), and logic. He had worked with ENIAC[6] during World War II and was a chief proponent of electronic computation during the infamous Manhattan Project.[7] von Neumann's seminal work in computer science was based throughout on the notion of brain-like construction.[8] Consider the following passage:

> The number of active organs [sic] in a large modern machine varies, according to type, from, say, 3,000, to, say, 30,000 [this was in the 1950s, when vacuum tubes were still ubiquitous]. Within this, the basic arithmetical operations are usually performed by one subassembly . . . the arithmetical "organ." In a large modern machine this organ consists, according to type, of approximately 300 to 2,000 active organs . . . certain aggregates of active organs are used to perform some memory functions. These comprise, typically, 200 to 2,000 active organs. . . . Finally, the . . . "memory" aggregates . . . require ancillary subassemblies of active organs, to service and administer them. . . . For all parts of the memory together, the corresponding requirements of ancillary active organs may amount to as much as 50 per cent of the entire machine.[9]

---

[6] ENIAC was constructed at the University of Pennsylvania in 1945. The machine was not even as powerful as a modern scientific calculator. von Neumann had a hand in further refinement of ENIAC's logic circuitry, this contribution being one of the very first examples of optimization.

[7] During the Manhattan Project (the creation of the first atomic bomb) there was, of course, very little electronic computing; yet von Neumann saw the tremendous power in such technology and explained the ideas vigorously to colleagues. It seems fair to say that, at least in this regard, von Neumann was one of the very first "evangelists" in the field.

[8] Karla von Neumann, wife of John, says as much in her preface to von Neumann 1958. Incidentally, this preface is so moving and yet so incisive that we believe any aspiring author should peruse it.

[9] von Neumann 1958.

This reads very much like an elementary anatomical theory of living brain structure. There are compartmentalized functions. Special organs support others. Note that by modern standards, von Neumann was not talking great numbers of parts. Perception of brain *anatomy,* rather than an eagerness for sheer quantity, was the original design factor for computing as we now know it.

WHEN THERE WAS ONE TRANSISTOR PER PERSON

Once the anatomical motif was laid out, the way was clear to expanding on parts within that framework. What would fairly consume the electronics industry from 1960 on was the explosion of semiconductor parts. The transistor, invented in 1945, is of course the original semiconductor amplifier. It is amusing to contemplate the question of when there existed in use, on average, one transistor per person. Well, the semiconductor industry itself can be said to have begun in the late 1950s, when discrete transistors seriously went to market.[10] By 1970, what with radios, digital clocks, digital watches, and some rudimentary modern calculators, there were billions of transistors in circulation, perhaps roughly one for each person on the planet. Such speculation is not a vacuous exercise, because it conveys an idea of scale and of expansion. The proliferation was extremely rapid: The transistor population went from zero to billions in less than one human generation.

The emergence during the 1960s of integrated circuits—arrays of interconnected transistors on one or a few substrates—was motivated somewhat by the need to put more parts into smaller spaces. One can also cite the incredible cost reduction of fabricating thousands of transistors, all at once, onto one "chip" (it was possible to fit this many onto one chip as early as 1970) compared with the cost of that many discrete transistors.[11] But whether the goal was cost-effectiveness or multiplicity of devices, the appliances and computers that emerged during the early days of integration universally enjoyed the advantages of, shall we admit, more parts.

---

[10] Malone 1995, pp. 53–54.

[11] At an astounding seminar in the 1970s, an engineer from a prominent semiconductor company lamented that digital watches could not be made any cheaper because, in spite of myriad transistors on the watch's internal chips, not to mention the quartz timing crystal, the single most costly part was the plastic wristband, whose manufacturing cost just couldn't be brought down any further.

Still, some things cannot be done with even unlimited numbers of transistors. There is a persistent belief that enough parts will "get the job done," and such beliefs fan the flames of the conspiracy. Certain computational hurdles are unassailable by dint of parts count alone. Consider first the total number of CPU cycles[12] performed across the entire history of computing. It turns out that this number is dominated by modern personal computers, because there are so many of them. For example, huge government supercomputers certainly generate many CPU cycles per computer, but there are on the order of 100 million personal computers in use in the world. The total number of cycles in history is difficult to estimate precisely, but it turns out that right around the year 2000 (or perhaps already, in the late 1990s) the total number will have reached the celebrated Avogadro number of chemistry. This number is about 6×10^{23}, or "6" followed by twenty-three zeros. Thus the total number of instructions is on the order of the number of carbon atoms in a little handful of elemental carbon. Any counting process or deep mathematical process (such as the factoring of a large number into two factors) that requires, say, 10^{40} instructions, is wildly beyond reach and would remain so even if all the transistors in all the computers on Earth were brought to bear on the calculation.

Again to convey an idea of scale, consider the total number of CPU operations required to render the completely synthetic movie feature *Toy Story*. This full-length movie required 10^{16} or 10^{17} machine operations, the greatest effort of its kind in history, yet this operation count is less than one-millionth of the Avogadro number.[13] Of course, it is not surprising that there are problems beyond the reach of any currently conceivable hardware/software configuration. But the effectiveness of more and more parts should not be thought of as asymptotically infinite. Then, too, advances on certain very hard problems (such as factoring) have been algorithmic enhancements—that is, they have been improvements in the methods rather than the hardware.[14] Perhaps the lesson of factoring will one day be

---

[12] CPU cycle means central processing unit cycle. A computer instruction, such as the addition of two numbers, can be thought of as a few CPU cycles, so we are essentially contemplating the total number of arithmetic instructions ever performed.

[13] *Toy Story* is a Pixar-Disney production. The rendering of surfaces, textures, and so on was performed at Pixar, using hundreds of workstations operating in parallel fashion.

[14] The algorithmic breakthrough phenomenon is exemplified in the 1991 factorization of a "most wanted" number called F9. Because of a new algorithm called the Number Field Sieve, F9 was demolished (all its factors were found) in a total of roughly 10^{18} CPU cycles. With previous algorithms alone, this number would probably still be out of reach. See Odlyzko 1995. This and related computational issues are discussed in Crandall 1997.

interpreted in such a proactive and universal way that the conspiracy of parts will be positively addressed, by way of a general algorithm revolution.

A GAME OF LEAPFROG

One of the more satisfactory technological developments of the five decades of modern computing has been the emergence of truly useful software. One class of useful programming software began with the interactive BASIC at Dartmouth in the 1960s. What a great idea (now commonplace, of course)—that a person type into a computer and get an immediate response on the terminal screen! Another milestone was the changeover, commonplace by the mid-1980s, from character-generating chips with their alphanumerical-only CRT displays to displays on which actual pixels are manipulated.[15] In fact, this innovation has deepened the conspiracy of parts in a subtle way. The new pixel-based displays gave rise to ever-more-complex software developments. A veritable explosion of graphics-related software in the 1980s has brought us, by now, such niceties as personalized, color-splashed World Wide Web sites, spectacular scientific visualization, and even full-length digitally rendered movies such as the revolutionary *Toy Story*. It should comes as no surprise that when the pixel-based personal computers emerged, there was a quantum leap in memory requirements. By the end of the 1980s, a personal computer worth its salt required at least a few megabytes of memory, a megabyte of memory being built from roughly ten million transistors. Whereas there was one transistor per person at some point in the 1960s, today the number has risen to more like millions per Earthling.

One might have thought that advances in software would somehow stem the parts expansion, by sharing—or at least refocusing—the computational labor. Instead, with few exceptions, the overwhelming effect of modern software has been to initiate a kind of leapfrog game. In this game, an incremental software development gives rise to a corresponding increase in hardware requirements. In turn, these improvements in hardware open the door for software that might not have run effectively on earlier hardware. One manifestation of this game is the common scenario in which a citizen obtains a new operating system, complete with some special new graphics options (such as animation) and is then forced also to

---

[15] Pixel, or *pel*, stands for "picture element." By 1980 it had become clear that personal computers would soon possess enough hardware horsepower to draw individual characters and figures, dot by dot—that is, pixel by pixel.

obtain an exotic color display of some sort. Of course, one responds, a color display for color software. But often in this circumstance, more memory must be purchased and more parts with which to support the denser color graphics display. In another manifestation of the game, any sufficiently new peripheral requires some kind of driving software. A new disk drive or sound device often necessitates new "driver" software that naturally takes up more space in memory, and so goes the leapfrog.

But there is no fundamental requirement for this kind of leapfrog coupling. The conspiracy of parts does not necessarily arise in every design system. For instance, systems of the natural world have in large measure rejected leapfrogging in favor of other growth and development schemes.

THE "COTASKING" OF BIOLOGICAL SYSTEMS

Certain biological systems exhibit what we shall call "cotasking." Cotasking is, in its most elemental form, a sharing of processing duties. A more exotic form of cotasking, which necessitates a long time scale, is coevolution. The simple aspect, process sharing, is to be contrasted with the phenomenon of one aspect of a system dragging the other into greater complexity or extent. A good example of cotasking is the manner in which the human ear and eye share responsibility. When one watches the slow vibration of an object, such as a speaker cone undulating subsonically at ten cycles per second, one cannot simultaneously *hear* such a low frequency. But it happens that as the frequency increases, the moment when one no longer sees the core vibrating but perceives only a blur is the moment when one begins to hear the vibration. The ear and eye probably coevolved in many respects, but at least in this one: There appears to have been no need to replicate resources for perception of important phenomena such as vibration. The eye and ear have coevolved a precise sharing of responsibility. For crude comparison, imagine procuring a digital camera peripheral and a microphone peripheral, both to be attached to a computer. Each peripheral will require its own software, and these software sets will be virtually independent. What is more, the new peripherals will act to displace each other in memory and to compete for overall system resources such as CPU cycles. The technological components are less compatible than the organic ones, partly because the human visual-auditory system has not developed on the principle of sheer addition of parts. Electronic and computer engineers have certainly exploited nature's peripheral designs; there is no question that modern digital cameras act "like the eye" or that high-fidelity microphones are physically reminiscent of the ear. But what has not been properly exploited is the notion of such peripherals working *together* and, in particular, working without undue displacement of each other's processing support.

Another elementary biological example of cotasking is the simple redundancy of bilateral symmetry. One hand can often do what the other cannot. Damage to one limb need not be incapacitating. This redundancy is exquisitely simple, but that simplicity should not deter us from seeking applications to computer design. Again, it is hard to find a nontrivial computer analog to bisymmetry, an analog that transcends the trivial expedient of employing, say, two computers. It would be good to see built-in redundancy in *one* computer that somehow avoids the materials cost of having two.[16]

It appears to be a fundamental principle of evolved organisms that parts often serve more than one purpose. Fatty tissue deposits play a crucial metabolic role and also insulate thermally. The cornea of the eye is an obvious environmental shield, and it also provides critical prefocusing, something like a self-cleaning transparent refractive camera cover. Examples abound, and one suspects that evolution is grounded in the phenomena of cotasking and redundancy. It cannot always be best to proliferate parts. Indeed, most mammals are of roughly the same "parts count," and yet the special functions of each species are varied and effective.

One might object that these simple examples of cotasking and redundancy are physical curiosities that have little to do with data processing. But there are more sophisticated examples that might lead directly to superior computer technology. One example of process sophistication involves the human visual system. It is now known that a certain amount of signal preprocessing occurs within the eye itself—that is, before the brain gets into the seeing act. The retina at the back of the eye can no longer be thought of merely as an assemblage of tiny cameras that together shuttle raw data to the brain. Consider the act of gazing at the full moon. If one does the physics, one finds that the bright circular spot actually cast upon the retina is quite small. In fact, on the order of only one hundred retinal

---

[16] There is, in fact, a minor example of redundancy correction in some computer designs. There are systems in which computer words are "parity-checked," which means roughly that the string of 1's and 0's is given a parity—something like the frank number of 1's involved. If this parity, a kind of signature of a string, is wrong, then an error is reported. Unfortunately, perhaps, this scheme is costly enough that in the modern heyday of parts proliferation, the parity option is often omitted.

cells are illuminated by that tiny image of the moon at the back of the eye. Yet no one will deny that the moon looks immaculately circular. By contrast, imagine a disk drawn on the screen of a personal computer and having a few hundred pixels. This will be a very sloppy disk indeed, with staircase effects on the edges. So where in the human system does the perceived moon image acquire its circular shape? It cannot be only in the brain that the hundred-pixel image is "corrected," even if the brain "knows" this is to be a perfect circle. For one thing, if a facsimile moon, constructed as a rough pixel image with its staircase edge effects, were put in the sky, it would not look like a perfect circle. Evidently, by some manner of signal averaging and preprocessing, the retinal cells convey to the brain a circular image. Today, the phenomenon is far from being understood. But we can say that in no existing camera-computer technology do the camera and processor have this tight symbiosis.[17]

NEURAL NETWORKS AND GENETIC ALGORITHMS

Two promising avenues of research that could conceivably dent the conspiracy of parts involve new system approaches. We refer to the study of neural networks and of genetic algorithms. Either of these new system approaches can, in principle, be manifest as a hardware or software design. By neural networks we do not mean living ones, although it is encouraging that neural network computing is modeled not on any idea of massive parts proliferation but on certain cotasking and redundancy features of living neural systems.[18] Similarly, genetic algorithms are modeled on prevailing notions of genetic evolution.

The neural network idea is not fundamentally difficult. Briefly, a nonlinear node—something like an adding machine but with nonlinear behavior—accumulates all of its inputs and produces a single output. Imagine a diagram with a set of input lines entering the node and the single output

---

[17] To belabor our central thesis further, one modern scheme for advanced visual reception runs as follows. In between the camera peripheral and the computer, one may insert a DSP (digital signal processor), a chip or chips to effect some rapid signal processing of the raw camera signals. In other words, one inserts parts into the manifestly *serial* processing chain.

[18] We caution that neural network models are not necessarily adequate for explaining actual living neural systems. Indeed, there are scholars of neurobiology who reject the notion that neural network computing will enhance our understanding of living counterparts. We speak instead of neural network computers as having inherited certain useful features of the true, living ones.

line having a value that is a complicated function of the inputs. The function performed by the node is based on a set of "weights" that basically determine the relative contribution of each input line. The intriguing conceptual "next step" is to contemplate many nodes, somehow interconnected. For example, the single outputs of four nodes could be the four input lines of a fifth node. This basic design is of course reminiscent of a living scheme of neuron interconnections. A typical application of a neural network computer is image recognition. Imagine that a camera, aimed at a printed letter A, feeds some set of signals (such as which regions of the image are printed black and which remain white) to an initial layer of nodes. Then perhaps a few more layers of nodes, ending with a final layer of 26 nodes (one for each letter of the alphabet), are all immediately energized in such a way that the final output line for A gives a pulse, the remaining 25 lines lying dormant. This means we have recognized the letter A. The difficult part of this image recognition design—as with all neural network models—is to arrive at the correct weights for the respective neuron connections. There are modern algorithms for self-training (the whole network is "rewarded" for more accuracy) and for such phenomena as back-propagation of error (mistakes are fed differentially back to the system of weights). But the tantalizing fact is that no matter how tedious the training of the network, once the letters of the alphabet are all "trained in" successfully, the network will perform its duties in a predictable, effective way.

Neural networks have potential for many types of recognition and processing that have proved difficult within the standard digital paradigm. Medical anomalies might be recognized from subtle X-ray perturbations. Complex economic problems, especially when the inputs are legion, might be solved. The stultifying problem of speech recognition may be assailable via neural network models. Eventual success in this latter endeavor would not be too surprising; indeed, yet another example of cotasking is the marvelous interplay between human hearing and vocalization. In fact, a truly comprehensive speech recognition system may be impossible to achieve without first understanding the interplay of voice and ear. What may one day emerge: neural networks that use common weights and connections both to listen and to speak.

An appealing facet of the neural network approach is that in two separate ways, neural network design is not exclusively parts-intensive. First, as we have intimated, the thorny problem in neural network design is not the accumulation of many nodes but the determination of weights. In terms of our crude biological analogy, it is not just the number of neurons that matters but also the way they are interconnected (and, of course, the chemical state of the system). In other words, we do not yet know how to train some neural networks that have a mere thousand nodes. One might say that the design problem is function-bound, or algorithm-bound. The second manner in which neural network concepts transcend the conspiracy of parts involves redundancy. It is possible to construct a neural network for which some subset of lines, or even nodes, can be rendered dysfunctional, and yet the overall performance is negligibly affected. This feature reflects the cooperation manifest in the nonlinear functions of many inputs at once and represents a form of what is called "majority logic." When the majority logic rules, small perturbations cannot carry the election.

Genetic algorithms, which have so far been applied mainly to software but can in principle be given hardware realizations, can be described as follows. Imagine a string of bits that together describe not a datum per se but a solution to a given problem. For example, the problem might be to find the highest mountain peak on the planet, and the desired string of bits—the solution—would hold the latitude and longitude of Mount Everest. To arrive at the correct string, one can imagine "parents" of, say, coordinates all over Asia, each parent being a coordinate string. Then a database (a digital map) is used to assess the height above sea level of each parent. Then take the highest ten percent of the parents, and kill off the other ninety percent. This is the selection stage, modeled on the notion of Darwinian natural selection. Now let every parent's bit string mutate; that is, flip a few of the bits in each surviving string. These new strings can be thought of as children, and to keep the population stable, we can have every parent beget say ten mutated children. Now we have a new generation of child strings, who become the parents of the next generation. We assess the heights of all the new parents, again apply the selection process, and so on. Success is expected after some number of generations: There will be a "best" child that, it is hoped, will be located, by virtue of its coordinate bits, right at Mount Everest.

Thus the genetic approach is to employ the notions of natural selection and mutation to "evolve" to a solution. The peak-finding example is not the final story, for there is current research into self-evolving programs and systems—sophisticated scenarios in which not just a quantity (such as height) is to be maximized, but an overall performance criterion is to be optimized. How is this promising approach related to the conspiracy of parts? The genetic approach is one of the few ideas that may apply *when the parameter space is overwhelmingly vast.* To exemplify this principle crudely, there may be no way to search an entire map database, point by point, in order to locate the highest mountain. But foraging with mutating organisms, where here mutation means trivial geographical displacement, may yield some success in statistical fashion. And a small population may suffice. If a database contains terabytes of entries, a genetic optimization algorithm might be effective when "brute force" search is not. In the future, then, the temptation to add memory or subsystems may be reduced by way of a superior genetic approach.

THE PROMISE OF NANOTECHNOLOGY

We have mentioned what are essentially algorithmic means of circumventing the parts conspiracy. There is, however, another branch of technology whose asymptotic limit—at least as imagined by the aficionados of the field—would shatter the conspiracy by head-on attack. We speak of nanotechnology and molecular nanotechnology, the fabrication of microscopic (even submicroscopic) engines that can themselves perform assembly. A thorough exposition of this deep and controversial field is beyond the scope of this essay.[19] Yet we can describe the features of the nanotechnological field necessary to formulate a conceptual attack on the conspiracy. The rudimentary idea is this: If nano-parts that replicate themselves allow the cost of parts to approach zero, why worry about the parts explosion? Forget about fancy statistical approaches for optimization with respect to a vast database, and just tell the nano-parts to replicate enough— say quadrillions—of themselves to effect an exhaustive solution to the computational problem.

It is a useful rule that in contemplating possible applications of nanotechnology, one should not be shy. When one speaks of miniaturization at the molecular level, many strange new constructs and functions become theoretically possible.[20] And when one speaks of automation at the mole-

---

[19] See Regis 1995 for an amusing and in some ways frightening exposition of the modern nanotechnological controversy.

[20] It is possible to imagine nano-device functions that border on the absurd. One of the authors once enjoyed a lively discussion with the inimitable Marvin Minsky on the then-popular controversy about what penalty (if any) should be inflicted for public burning of the American flag. The idea came up that, certainly, a very stiff penalty should apply if one engineered "flag powder," consisting of nano-flags with long molecules for the stripes and alkali metal atoms for the stars, so that a handful of the powder (such a macroscopic parcel might appear violet) could be flashed at the touch of a match. The criminal charge for a single such flash would presumably be something like 10^{20} federal counts of flag desecration!

cular level (for instance, a factory built of nano-parts that in turn manufactures nano-parts), the limits are hard to imagine.

Let us explore some potentially practical applications of nanotechnology. Imagine a "dynamic painting" as a large rectangular framed region on a wall. A small tray, like a chalk tray, would contain brightly colored "nano-pixels." These pixels would be nano-structures, each containing some complicated receiving and locomotive machinery, but each also adorned with a bright primary color: cyan, magenta, or yellow. A remote computer could then display a whole painting by instructing each nano-pixel to move to a certain unique coordinate location on the painting and sit there. Except for color identity (cyan, magenta, yellow) and perhaps a unique "password" for receiving movement instructions, each nano-pixel could be identical to all others.[21] One can envision a future in which printing devices of all types are equipped with such "nano-ink." Given the infinitesimal cost of nano-pixels, say the nanotechnology devotees, such printing would not be expensive. There is an interesting cultural point: After printing, the nano-pixels in place on the medium would be powered-down, say forever. Given a little care on the materials engineering side, such ink would presumably last for eons. One might even envision "recyclable nano-paint" that would, along with "nano-paper" (assembled, of course, from nano-fibers that self-assemble into a fabric) allow subsequent, instantaneous reassembly into fresh material.

Nanotechnology could also be used in the field of medicine. One envisions a flotilla of nano-repair-vehicles in the human bloodstream, attacking microscopically what the natural immune system cannot or ferreting out and even attacking microtumors. In the true spirit of molecular nanotechnology, the blood nano-machines could self-delegate and self-maintain, much as the components of the natural immune system do.

In the computation arena per se, nanotechnology could presumably blunt the conspiracy of parts by allowing constructs such as memory (even dynamically replicating memory) to expand at infinitesimal cost. One might envision a terabyte memory system for which any attempt to overrun memory with excess data would be met with a virtually instantaneous

---

[21] With identical nano-pixels, the remote computer signal could, for example, demand the correct color at each location, and the pixels could motor and shuttle around until they arrived statistically at the correct ensemble for the painting. The assembly of every new painting would probably look like the reversal of a movie of the picture melting into swirls and rivulets of paint.

addition, through self-replication of memory nano-cells, of another giga-byte.[22]

Strangely enough, there is no precise biological analogy for the nan-otechnological solution to the conspiracy of parts. Biological parts regen-erate and self-organize, of course, but they do this under stringent regula-tory mechanisms that are not fully understood. A frightening prospect is the possibility that nano-systems could expand for no reason. By contrast, the human brain volume does not just multiply at will, whatever the mo-tive to think or act more strenuously. There have been various pronounce-ments of the threat of "malignant" nanotechnology, exemplified by run-away self-replication that could turn the whole planet into "gray goo," a sludge of nano-machines still frantically groping for more replication ma-terial.[23] As horrific as this threat seems, we feel that except possibly for the issue of time scale (for we tend to imagine the threat as a problem for fu-ture generations), it is no more horrifying *in principle* than the present-day conspiracy itself.

Quantum Computation

One very promising area of research—primarily theoretical research at this point, because nobody knows how to fabricate suitable parts—faces head on the fundamental issue of machine design and, in so doing, shows some potential for future containment of the conspiracy. This is the field of "quantum computation," in which the classical notion of a computing engine is challenged at the deepest level.

Consider first the classical Turing Machine, or TM, named after the great computer scientist Alan Turing. The TM is the general abstract model of all working digital computers. One may think of a TM as pos-sessing an idealized tape of instructions read by a processing unit. The unit's instructions might be such as "jump ahead to tape position X" or "place the contents of tape position Y into memory location Z" and so on. In other words, *every* conventional digital computer—complete with CPU (tape reader), software (tape), and memory—is of the TM class. As uni-versal as the TM concept is, there are problems whose solutions still re-

---

[22] The physical limitations are minor. Consider that 10^{20} molecules of decently complex structure still make up only a handful, and then imagine a lifetime of desktop computing having added a barely visible complement—just a speck—of extra terabytes to one's dy-namic nano-memory.

[23] Regis 1995, pp. 121–124.

A Network Orange

quire untoward amounts of time.[24] But it was in the field of computational physics that a non-TM design arose. In 1982 the eminent theoretician Richard Feynman noted that attempts to model certain physical processes with a TM would suffer from exponential slowdown. He proposed a quantum-physics-based computer that would handle certain physics problems with more ease than a TM.[25] Later, a more general model of what is now called a Quantum Turing Machine (QTM) was established, though a practical QTM has never been built.[26]

It has been shown theoretically that a QTM could handle certain problems in less time than a conventional TM consumes. This is possible because a TM is based on classical physics, where every state of the computing engine (such as current tape position and current instruction) is completely deterministic and well-defined. But in quantum physics, states are statistical, not completely defined. At first it would seem that any quantum-world computer would make mistakes. The twist, though, is that a virtual infinity of quantum states can be processed simultaneously. In the words of one of the pioneers of the field of quantum computation, "Quantum theory is a theory of parallel interfering universes."[27] Without getting too far into quantum-theoretical terminology, we can say that a QTM performs massive parallel processing by allowing quantum states to evolve naturally. Under certain mathematical rules involving randomness and superposition—rules believed to be absolutely true of the quantum world—an ensemble of quantum processing cells can work together to yield a result by way of quantum interference.

One (oversimplified) way to imagine a quantum computer is to envision a line of atoms, each set up in a particular state corresponding to a digitized data set. Then we allow this atomic line to evolve according to the natural laws of physics. That is, we simply prepare the atomic system and then "let it go." Under the rules of quantum theory, the atomic states at a later time might represent a "spectrum" of the original data set. The

---

[24] Problems such as large-number arithmetic, factoring of large numbers, physics simulations, and various difficult problems mentioned in Essay 2 are notoriously difficult for our conventional TMs.

[25] In Feynman 1982, a "spin-flip" quantum engine is proposed for modeling quantum ensembles for which a direct TM approach would be painfully slow. This idea has evolved into the modern notion of the QTM.

[26] The modern theory of QTMs extends, after Feynman's work, from the work of Deutsch (1985) to modern references such as Bernstein and Vazirani 1993.

[27] Deutsch 1985.

standard TM approach is to perform a fast Fourier transform (FFT) on the original data to get the spectrum. As is well known to computer scientists, an FFT must be computed holistically (one might say holographically) in the sense that each output datum depends on *all* the input data. The shuffling of all the input data in the conventional TM motif is computationally expensive. But the interfering quantum wave functions of the atomic sites already enjoy, by each sensing all the others according to the tenets of quantum theory, a certain holism. It is as though an FFT spectrum can be calculated all at once—nature's way.

Another way to grasp at least some of the principles of quantum computation—again we simplify—is to observe that laser light, in creating, say, a hologram of a solid object, performs a virtually immediate "wave function computation." About one-billionth of a second after the laser light strikes the solid object, a hologram, essentially a massive spectral transform, is generated at a camera's film plane. By contrast, a very fast modern computer (a TM, of course) would require perhaps one second or more to calculate the hologram. Thus they differ in performance by a factor of a billion. The interfering wave functions of the laser light calculate the hologram spectrum—nature's way.

By combining the notions of apparati and waves, one can go a little further. It turns out to be possible, via an assembly of polaroid filters and mirrors, to construct an apparatus that performs simultaneous addition and subtraction. Imagine a "black box" containing such optical apparati, such that if a wave of right-polarized light (of amplitude A, say) and a wave of right-polarized light (of amplitude B) are impingent, then out of the box emanate waves of left-polarized and right-polarized light with amplitudes $A + B$ and $A - B$, respectively. It does not sound like that much, to have the sum and difference emanate simultaneously. But the black box amounts to what quantum theorists call a "unitary transformation." And it turns out that massively complex calculations can be built up from such unitary transformations. It is possible to graduate, in principle, to much more algebra than $A \pm B$ and, in fact, to handle many operations on many variables all at once, in, again, nature's way. Such schemes have been proposed for calculating the aforementioned spectra of signals and for factoring large numbers, all in genuinely far less time than the conventional TM would require. This waves-impingent-on-black-box notion strikes very close to the heart of the current QTM theory. In fact, thinking of the impingent light waves as particle- or state-based (that is, thinking of light as photons and of the black box as an operator for photon statistics) leads one to the full exposition of the modern theory. The QTM concept is a deep one, and to grasp it fully, it is necessary to confront the mathematical—even the philosophical—apparatus of quantum theory.[28] Lay-oriented review works do exist, however, that have evolved along with improved understanding of the QTM notion.[29]

The profundity of the subject is not surprising: A standard TM is so very general that *something* strange must emerge if we are to graduate to a new level of computing machine. Everything from an abacus to a cash register and from a home microprocessor appliance workstation to a supercomputing mainframe is a TM. Though a QTM has not yet been built, it is not unrealistic to think of such a machine as a small parcel of the universe, where the quantum coefficients are known and controlled. In this way, the parallel processing power of matter itself would be brought to bear on the conspiracy of parts. Here's a fanciful way of saying all this: To thwart the conspiracy of parts, why not use nature's own parts, the quanta all around us?[30] It is intriguing that future computer "hardware" might be fabricated via the nanotechnology we postulated in the previous section and yet involve nature's very own quantum interactions in the "software."

THE FATE OF THE CONSPIRACY

Finally, let us speculate on what may become of the conspiracy of parts over the coming decades. We summarize the conspiracy as follows: The trend for solving problems via more, faster, and smaller parts, coupled with the costly leapfrog game between software and hardware, has led to a common belief is that this runaway expansion is the correct way to effect new solutions. The computer makers are preoccupied with design rather than purpose. Ultrafast number crunching is too often assumed to be advanced number crunching. The working of many parts is often considered, despite myriad counterexamples, more sophisticated than the working of just a few parts. It is as though the result of machinations has been

---

[28] An excellent review article is Ekert and Jozsa 1996, in which both the theory of unitary transformations and the current thinking on the construction of actual quantum computer gates are expounded.

[29] A good starting point for the interested reader is Williams and Clearwater 1997.

[30] The quantum world not only performs natural processing but also exhibits natural memory. Anyone who has seen the glory of atomic spectra (colored spectral lines produced via a prism or grating) has witnessed such memory (the storage of energy quanta at consistent levels).

overshadowed by the machinations themselves. In view of our sure knowledge that modern computing is a direct consequence of world war, we cannot help but take seriously the insight of Norbert Wiener:

> There is one quantity more important than "know-how" and we cannot accuse the United States of any undue amount of it. This is "know-what" by which we determine not only how to accomplish our purposes, but what our purposes are to be. I can distinguish between the two by an example. Some years ago, a prominent American engineer bought an expensive player-piano. It became clear after a week or two that this purchase did not correspond to any particular interest in the music played by the piano but rather to an overwhelming interest in the piano mechanism. For this gentleman, the player-piano was not a means of producing music, but a means of giving some inventor the chance of showing how skillful he was at overcoming certain difficulties in the production of music. This is an estimable attitude in a second-year high-school student. How estimable it is in one of those on whom the whole cultural future of the country depends, I leave to the reader.[31]

The parts expansion might degenerate into a nanotechnology enterprise for which parts themselves are attained at vanishing cost. Though the *economic* dilemma may have been resolved, the ensuing blizzard of parts and systems might give rise to unimaginable technological horrors. So this nanotechnology limit—even if myriad parts behave in a highly sophisticated, "interesting" fashion—might not offer a *cultural* resolution to the conspiracy.

Another available path is to invoke the biological analogy mentioned previously. Neural network computers are not inherently parts-bound, given their reliance on training and capability for natural redundancy as opposed to brute force back-up redundancy. Incidentally, computer scientists may object that analog computing has proved costly and inaccurate over its many rebirths and failures, so whatever way we find to circumvent the conspiracy, that way must be expressly digital. We are not contradicting such sentiment. By neural network solutions, we mean expressly digital ones.[32] We are saying simply that a deep attention to the interconnection of parts might stem the parts expansion.

---

[31] Wiener 1954, p. 183.

[32] Technically, the weight coefficients for each node are thought of as high-precision floating-point numbers, as are the input and output levels themselves. In fact, without loss of generality, the output lines can all be single-bit or integer values, and so on. This design allows for a patently deterministic, digital (not analog) neural system.

A Network Orange

We believe that circumventing the conspiracy of parts in any way other than by the nanotechnology asymptote will necessarily involve adaptive machinery that encompasses various grand features of biological evolution. One potentially formidable synthesis would be the marriage of nanotechnology with quantum computation. Imagine a nano-quantum computer as "ultrabiological" in that computations would proceed unthinkably fast, with memory expanding effortlessly to unprecedented levels. Such a synthesis would amount to tapping the intrinsic organizational prowess of matter and energy in ways that living systems have never done. A nano-QTM marriage might be envisioned then as a parcel of the universe that, under nature's fundamental laws, does a programmer's bidding.

We also wonder whether genetic algorithms will become the staple of personal, business, or scientific supercomputing, or whether machinery has some mode of self-maintenance and dynamical internal reconnection. It would be refreshing to see software platforms undergoing *algorithmically* based rather than commercial selection and evolution.

In closing, we state what we take to be obvious: Living systems had to extrude themselves through the time line of evolution, whereas computing machines have not had to do this on anything like an evolutionary time scale. As a result of this limitation, we suffer the conspiracy of parts, a conspiracy of stark constructionism, a grand technological effort that mirrors the wonders of living systems only in a spirit of caricature.

REFERENCES

Augarten, S. 1996. *The Invention of Personal Computing.* Unpublished manuscript.

Bernstein, E., and Vazirani, U. 1993. "Quantum complexity theory." *Proc. 25th ACM Symp. on Theory of Computation,* pp. 11–20.

Crandall, R. E. 1997. "The challenge of large numbers." *Scientific American,* February, pp. 58–62.

Deutsch, D. 1985. "Quantum theory, the Church-Turing principle and the universal quantum computer." *Proceedings of the Royal Society of London.* A400, pp. 97–117.

Ekert, A., and Jozsa, R. 1996. "Quantum computation and Schor's factoring algorithm." *Reviews of Modern Physics* 68(3), pp. 733–753.

Feynman, R. 1982. "Simulating physics with computers." *International Journal of Theoretical Physics* 21(6/7), pp. 467–488.

Kauffman, S. 1972. *Towards a Theoretical Biology.* New York: Oxford University Press.

Landauer, T. K. 1995. *The Trouble with Computers.* Cambridge, MA: M.I.T. Press.

Malone, M. S. 1995. *The Microprocessor: A Biography.* New York: TELOS/Springer-Verlag.

Odlyzko, A. 1995. "The future of integer factorization." *Cryptobytes* (technical newsletter of RSA Laboratories).

Regis, E. 1995. *Nano.* Boston: Little, Brown.

Varga, R. 1996. "Obituaries: Olga Taussky Todd," *SIAM News* 29(1), p. 6.

von Neumann, J. 1958. *The Computer and the Brain.* New Haven, CT: Yale University Press.

Wiener, N. 1954. *The Human Use of Human Beings.* New York: Da Capo Press.

Williams, C. P., and Clearwater, S. H. 1997. *Explorations in Quantum Computing.* New York: TELOS/Springer-Verlag.

TOWARD A THEORY OF MACHINE CONSCIOUSNESS

T he superior [chess] player is always on the alert for exceptions to general rules and analysis. . . . One of [Alexander] Alekhine's particular strengths lay in his ability to create positions that on the surface seemed rather unclear or innocuous, but in fact were deadly. Viennese Grandmaster Rudolf Spielmann complained that he could understand and discover Alekhine's final combinations easily enough . . . but that in his own games he was unable to *develop* the kinds of positions which Alekhine concocted. Since computer chess is so much based on "normal relative values" in its evaluation functions, skeptics will wonder with good reason how computers will ever reach expert or master strength.[1]

THE BOONDOGGLE OF ARTIFICIAL INTELLIGENCE

At the threshold of a new century, chess computers are nipping at the heels of grandmaster-level play. Soon enough the machines will systematically demolish human grandmasters. (In some isolated cases the demolition has already occurred.) However, much of the advancement in chess

---

[1]E. Hearst in Frey 1983, p. 189.

machinery over the last two decades is of the "brute force" variety. Machines can now sustain enough MIPS and megaflops to overwhelm, via wide sampling of positional alternatives on every move and libraries of move sequences, virtually any chess player, and certainly the average chess player. True, there have been notable enhancements in chess software. But one wonders whether any of these improvements has cast sufficient light on questions like Spielmann's, implicit in the quote above: What is the true difference between normal play and expert play?[2] It may be that the issue is somewhat resolved for chess play per se, but certainly the modern tendency to adopt brute force solutions has not shed much light on the general problem of "normal relative values" for other artificial intelligence (AI) dilemmas.[3]

One motivation for the development of a theory of machine consciousness is the continuing failure of AI to produce goods of practical value. For most denizens of the civilized world today, all AI has brought to the party is something like microwave appliances that "usually" do something well. As Dreyfus put it,

> For now that the twentieth century is drawing to a close, it is becoming clear that one of the great dreams of the century is ending too. Almost half a century ago computer pioneer Alan Turing suggested that a high-speed digital computer, programmed with rules and facts, might exhibit intelligent behavior. . . . After fifty years of effort, however, it is now clear to all but a few diehards that this attempt to produce general intelligence has failed . . . it has turned out that, for the time being at least, the research program based on the assumption that human beings produce intelligence using facts and rules has reached a dead end, and there is no reason to think it could ever succeed. Indeed . . . Good Old-Fashioned AI (GOFAI) is a paradigm case of what philosophers of science call a degenerating research program.[4]

---

[2] In 1996 world champion G. Kasparov won a tournament, by a disturbingly slim margin (2 wins–1 loss–4 draws), against the IBM machine "Deep Blue." This epic struggle is described in Newborn 1996. More recently, Deep Blue beat Kasparov in a rematch (May 1997). Some profound issues arose in this second match: (1) Kasparov perhaps rightly criticized the IBM programmers' lack of candor in not supplying previous Deep Blue game listings. (2) Deep Blue made good use of stored games, moves, and positions. (3) Kasparov's behavior near the end of the match suggests that in such confrontations, a machine can sometimes be the better sport.

[3] The typical press statements concerning the obvious strength of Deep Blue have revolved around how *fast* and *extensive* the hardware is. Certainly, the software programmers on the IBM side deserve due credit, but the point is that evidently the modern way to convey this kind of machine excellence to the public mind is to conjure images of brute force.

[4] Dreyfus 1993, p. ix.

A Network Orange

A canonical example of a domain that AI has failed to conquer is speech recognition. We do not have readily available machines that recognize speech at a satisfactory level. Even the most advanced research manifestations of speech recognition are faulty in obvious ways; with few exceptions, they are error-prone, and they are never contextually sharp. No machines can carry on the legitimate conversation of which a five-year-old human is quite capable. The phenomenon that surprised and fascinated Weizenbaum in the 1970s, that the original conversation-generator ELIZA was enthusiastically embraced by many as a valid converser,[5] still surprises and fascinates. One does not have to look far to see someone acting with marked deference toward a very minor example of AI, as though the machinery were endowed with some kind of emergent vitality. Regardless of a machine's power or lack thereof, humans rush to bestow on it certain emergent features that are not, in fact, present. We suspect that this tendency arises from a simple connection: All computing machinery is built on some sort of model of how we think the human brain functions. It will turn out that this simple idea, especially in regard to the problems of input and output, is a cornerstone in the construction of a possible theory of machine consciousness.

On the matter of AI for the common person, we seem to have learned more about human fantasy and longing in regard to what would be ideal machines than we have learned about machines themselves. In this sense, AI has been a kind of boondoggle, a scientific-social-political program without beneficiaries. Still, one can only respect the various AI research efforts that do not now, but may one day, benefit the masses. This in turn provides a second motivation for the development of a theory of machine consciousness: Much of AI research may be moving in the wrong direction. Rather than merely building on the assumption that machinery is slightly smart and needs to be made truly smart, AI research should incorporate a theory of machine consciousness.

DOUBLE OBFUSCATION

The very phrase *artificial intelligence* is a double obfuscation. Not only do we not know what an intelligent machine is; we do not know what intelligence is itself. It is not even clear to what extent a valid definition of intelligence should encompass behavioral phenomena. Is a strong chess

---

[5] Weizenbaum 1976.

machine, for instance, really playing chess? What about time pressure and other psychological effects?[6] Then there is what might be called "metastrategy," in which one chess player intentionally invokes a singular variation on standard play; the variation may not even be fundamentally sound, but it may cause the opponent to become disoriented and therefore defeatable. Although chess does admit of mathematically rigorous, programmable, deterministic rules, we do not think these cultural aspects are frivolous. Indeed, in a phrase such as *artificially intelligent chess machine,* it seems the only properly defined word is *machine.* Unfortunately, the normal public interpretation of the phrase is too automatic and innocent. Those who imagine that brute machine force could settle chess matches mathematically exhaustively—and so obviate the need for any definition of *intelligent*—might contemplate the staggering number of possible chess games, which is something (very roughly) like 10^A, where A is the gargantuan Avogadro number 10^{23}.[7]

EXTREME DIFFICULTY

Partly to dispel any notion that we are bent on sheer criticism of current AI research, we shall explore the extreme difficulty of various problems that plague the field. Major barriers lie in the path to effective machine intelligence. In what follows, we shall concentrate not just on levels of difficulty but also on *why* this or that problem is so difficult.

First, let us once again discuss the world of chess. Even though machines are virtually at the grandmaster level, severe algorithmic problems remain. Two older but telling examples from the software programmer's folklore are "Lloyd squares" and "Fischer regions,"[8] both of which notions

---

[6] Tales of chess psychology abound. World champion Mikhail Tal was infamous for his daring sacrificial combinations that, even when mathematically unsound, involved notable psychological content, which was enhanced by his intense over-the-board stare. On a somewhat more primitive level, the ancient master Ruy Lopez once recommended that a player should endeavor to place the chessboard so that reflected room light would glare into the opponent's eyes.

[7] Readers who are still willing to embrace the notion of an "artificially intelligent chess machine" may wish to contemplate the bleak asymptotic terminus of chess machine development. Imagine two galaxy-sized supercomputers squaring off: The white machine moves pawn-to-king-four, and after some light-travel delay, the black machine resigns. Would we not hesitate to call that chess?

[8] A Lloyd square, named after the great puzzlist, is a square that is in some sense the least attractive place to move yet turns out to be the key to a problematic chess position. A Fischer region, named after the inimitable Bobby Fischer, is a region (in play space) of future move combinations that is very difficult to infer from the current position yet is where the

are virtually impossible to program, especially if all one does to improve chess software is increase machine horsepower and perhaps, via memory or bus expansion, inflate the overall play space. (Play space can be thought of as the set of possible chessboard configurations for some number of moves into the future—a kind of higher-dimensional chessboard.) One defect of many chess programs is an obsession, in effect, with locality in play space. We must always remember that the machine does not necessarily ponder a wide region of play space, but perhaps just the currently most interesting regions of conflict. This myopia is especially prevalent in older chess software for which most of the algorithmic effort goes into tactics—that is, into local combinations confined to the relative hot-spots in play space.[9]

A second AI problem of extreme difficulty is speech recognition. The goal is for computing machinery to "understand" actual vocal speech. This goal is very much harder than the one set out for the ELIZA system, for example, in which the human subject types in English sentences. The speech is to be spoken into an input peripheral (microphone); then the computing machinery, ideally in real time,[10] parses the speech. Even the lesser problem of having the computer merely print out a text of such speech is frightfully hard. The extension of responding to or acting on such speech is even harder. Rather than delve into the detailed technology of speech recognition, we present an excerpt from a modern research treatise. The following paragraph conveys an almost humorous notion of just how intricate this technology has become.

> As a first step, we compared the performance of three recently developed word boundary detection algorithms to an algorithm . . . based on energy levels and durations, which is enhanced by automatic threshold setting. . . . We report their performances when integrated with a commonly used speech recognizer, vector quantization-based hidden Markov model (VQ-based HMM). The VQ-based recognizer used first

---

decisive tactics will unfold. Frey (1983, pp. 131–132) cites an instance where Fischer himself discovered a profound, nontrivial continuation in a middlegame against Viktor Korchnoi. And this was in five-minute speed chess, no less.

[9] For example, many commercial chess programs can be beaten in the following way. A long-term attack on the machine's castled King position may not raise the machine's internal alarm until it is too late to defend against your slow attack. Minor pieces and pawns, perhaps, can be brought to bear down on the machine's King, but with such delay and finesse that the program remains oblivious to the attack until it is too late.

[10] In this case, *in real time* means "as the speech comes in," so the machinery never lags behind.

and second order regression features . . . extracted from the index weighted cepstral coefficients derived from the twelfth order model of perceptually based linear prediction analysis. . . . The training was done on clean speech produced in a normal environment (without background noise) and the testing on Lombard or noisy-Lombard speech. Accuracy was judged by recognition rates.[11]

It is hard to imagine a more convoluted and detailed passage, even on a technological research subject. Yet this paragraph is indicative of just how difficult is speech recognition. If Richard Feynman was right when he said that one does not understand a difficult subject until one can put the explanation in simple terms, then we certainly do not understand speech recognition.

Beyond the technical difficulties of deciding whether to use a Markov model or to assume noisy-Lombard speech, there is the awesome problem of context. Again, we do not want to get into a protracted discussion of semantics and context, but we shall mention some key aspects of the context problem. There are at least three subproblems, which we can call the audio-context, environmental-context, and spelling-context problems.

The audio-context problem concerns the heightened accuracy of the natural human speech recognition apparatus when the source is identified. A good example is the incompletely understood "cocktail party effect," in which one may understand the speech of one party goer, even over a surrounding ambient din. To be truly amazed by this focusing skill, one need only try to computer-analyze party speech recorded by a microphone. It is extremely difficult to computer-process speech by *one* loud and clear speaker, let alone a speaker whose words are buried in ambient talking. Another manifestation of the audio-context problem is the detection of silence. For a computer to bracket correctly a spoken word in time, it must decide precisely when the word stops and starts. But this is (literally!) much easier said than done: Humans have an uncanny ability to know when speech starts and stops, even when massive noise is present. Imagine, for example, the spoken phrase "disease ends." Especially in the presence of noise, the final "s" sounds trail off into such noise, so that faulty silence detection might yield the machine-perceived result "dizzy

---

[11] The passage is from Junqua *et al.* 1994. Our intent is not to criticize their research paper, which is in fact a good paper, but to give a feeling for the vocabulary attendant on the state of the art.

end." Silence detection by machine remains an elusive research problem that is not yet satisfactorily solved.

The environmental dependence of human speech processing is difficult to quantify. A person in a closed room saying "I painted my walls" may well be understood even if the word *walls* is garbled, because the walls are present and obvious. The subtlest odor of new paint might allow even the word *painted* to be garbled and still be recognized—a good example of the interplay between natural anatomical parts. There are countless examples of these symbiotic human skills, and speech recognition research should ultimately take environmental context into account.

The spelling-context problem concerns the task of speaking printed speech, say by training a camera peripheral on a page, or accessing a text file, and expecting the computer to talk.[12] Though this is easier than recognizing spoken speech, it remains a difficult machine task. Many languages, including English, allow ambiguous spelling where multiple meanings can be unraveled only on the basis of context, and there are situations when the context is not even evident in the text itself.[13]

A third extremely difficult task is weather prediction. It is well known that today, supercomputers can fairly successfully be brought to bear on the hydrodynamical equations that govern air flow in Earth's atmosphere. But—and here most readers will need no convincing—weather computations are fallible. Instead of concentrating on the quantitative incidence of imperfection, we shall discuss something much more important: the reasons for it.

There are at least two reasons why computerized weather prediction falls short of omniscience. First, there is the fundamental problem of chaos. In this instance, chaos has a strict meaning: A microscopic perturbation may have ultimately macroscopic effect. There are many picturesque examples of physical chaos, all of which exhibit an extreme sensitivity to initial conditions. Another view is that in chaotic systems,

---

[12] We presuppose here that the "character recognition problem" is solved; that is, the letter images as gathered by the camera are turned by the computer into the correct letters. Though not quite perfect, such character systems do exist today and constitute one of the scant successes of AI.

[13] A fanciful example of an intractable spelling-context problem is the following statement: "Although a lightning storm forced an early end to the outdoor labor convention, all the hard-hat workers went home *unionized.*" (Did they form their desired union or did the lightning miss their metal helmets?) This kind of incomplete context may be rare, but our point is that any comprehensive software/hardware solution must be "wary" of contextual crises.

information about motion and behavior decreases (that is, gains entropy) at a significant rate as time passes.[14] Thus, for example, solutions to hydrodynamical equations may possess a great deal of sensitivity to initial conditions (input data for a weather system). Another way to put this is to say that vast machine power is required to track a microscopic disturbance whose destiny is a macroscopic effect.

Another difficulty in weather prediction is insufficient physical input. Everyone has seen real-time media tracking of hurricanes, wherein it is clear that in spite of all the crisp satellite photos and accurate wind speeds, nobody knows which way the hurricane will meander. On the basis of hydrodynamics, it seems the only way to predict hurricane direction would be to have much more *input* data than are currently gathered. One might even need vast banks of various sensors near the hurricane itself, or perhaps even in the eye of the hurricane. The best weather computers possess neither sufficient input data nor enough input peripherals. We shall have more to say about input data later in this chapter.

The science of earthquake prediction (or general cataclysm prediction, for that matter) likewise suffers from paucity of input. It is often said that seismologists just don't have enough input data, when what is really meant is that we do not understand the proper parameter space for the problem. Instead of contemplating banks of physical sensors for weather input (and assuming that the hydrodynamical model of weather is solvable), we contemplate instead a kind of "software input"—parameters that might not even need physical input peripherals but certainly must be understood. For example, if the moon's role in earthquakes were entirely understood, then the moon's gravitational-pull vector would properly belong to the seismographic parameter space. Then the "input values" for the moon would have to be obtained during computer prediction runs, perhaps with the aid of a database similar to a table of water tides. And who knows how many other parametric inputs should support earthquake prediction? Cataclysm prediction is hard, the very problem of data input being both qualitatively and quantitatively difficult.

A fifth problem is that of natural language generation—the creation of natural, meaningful prose, poetry, and music. As we mentioned previously

---

[14] An exemplary chaos metaphor is that of the butterfly in China that, by beating its wings, eventually causes a tornado in, say, Kansas. It turns out that this "butterfly effect" cannot happen as stated, but the metaphor illustrates the chaos concept. Furthermore, there *are* real effects on real weather when the butterfly is replaced by, say, a sudden heating or cooling over a large enough area.

in regard to the speech and printed-word recognition problems, there are semantic difficulties such as the context problem. With computer prose, one faces a converse of the context problem: What does the machine intend?[15] As we shall see later in this essay, a natural language generator might appear intelligent, but as of today, we can readily identify acute defects in such "intelligent" behavior.

The five AI problems we have touched on (algorithmic problems in programming chess play and the difficulties we encounter as we instruct the computer in speech recognition, weather forecasting, earthquake prediction, and natural language generation) amount to the proverbial drop in the bucket: AI problems are legion. We gave these five examples to convey a sense of how very difficult such problems can be. Moreover, we submit that all the problems have something important in common. Not only processing power but also severe I/O[16] difficulties stand in the way of machine intelligence, and thus of machine consciousness.

Progress in AI

Before we turn to I/O issues, it is valuable to note some difficult modern problems in AI for which there has been at least some success.

One example of successful solution is that computers can now talk pretty well, even if they cannot yet listen or read well. This is a prime example of output being fundamentally easier than input. Another area where some promise has been realized concerns learning. Parallel to modern neurobiological discoveries are various algorithms for machine learning.[17] There are good examples of progress in the general field of robotics. An era of simple household robots is imminent, if one can judge from the current state of the art of robot vision and execution.[18] The most striking achievements, however, are those where output behavior is useful or interesting. The deeper aspects of robotics, such as

---

[15] There are countless examples of converse context crisis. For instance, if "fruit flies like a banana" emanates from a machine, the context is absolutely unclear.

[16] I/O stands for input/output. The processes of procuring, causing, arranging, and specifying input/output, and even the hardware devices involved, have all been given the nickname I/O.

[17] There are even machine-learning models based on analogy to secreted chemicals such as neurotransmitters and hormones. One reference is S. Kitano's "A Model for Hormonal Modulation of Learning," pp. 532–538 in Volume I of Mellish 1995.

[18] Mellish 1995 is replete with descriptions of working simple robots, robot paradigms, and intelligent agents in general.

autonomy, intent, conditional behavior, and so on, still seem to be out of reach.

INPUT STARVATION

It is not true that most so-called artificial intelligence problems remain unsolved merely because "computers do not yet have enough processing power." The idea that CPUs do not wield sufficient power, say in terms of megaflops,[19] is a dangerous delusion, especially given the rush to more machine complexity. Our claim is that one cannot separate the boondoggle of AI from the "I/O problems" of AI. We have already examined cases in which not enough quality input is available to the (possibly very powerful) processor. What is more, there are cases, such as the failure of seismological prediction, in which one suspects input parameter starvation. Insufficient input and narrow input parameter space can be collectively called "input starvation." To amplify the input starvation problem, let us turn to some biological examples of I/O *par excellence.*

One reason why the human ear (for our purposes, think of it as an input peripheral) works together so exquisitely with the brain is that the ear's dynamic range is enormous.[20] What is more, the whole system has processing modes optimized for whatever part of the dynamic scale is prevalent. At the threshold of hearing—imagine the faintest conceivable whisper—the physical eardrum vibrates through something like one atomic diameter. Clearly, this is a design that works down to the minutest level. How the brain manages to hear sounds near the silence threshold, and filter these from noise, is not understood. To conceive of the equivalent signal-processing problem, one would need to imagine a design in which a great deal of the processing parts were dedicated merely to noise filtering. After all, at the atomic level, the eardrum is flapping to and fro over thousands of atomic diameters: The quiet whisper just causes some

---

[19] A megaflop is one million numerical (floating-point) operations, though the term is often used to mean the *rate* of one million such operations per second. We recall an interesting statement made in the early 1980s by a prominent plasma physicist, to the effect that the physicist's whole life had changed for the better "when 25-megaflop machines became affordable." Today, gigaflop machines (one billion operations per second) are commonplace, and one may dream reasonably of teraflop (one trillion operations) or even pentaflop (one quadrillion operations) machines or machine networks.

[20] Dynamic range is a measure of how different the softest signals (the faintest conceivable whisper) are from the loudest (jet engines, the threshold of pain). For the human ear, the ratio of loudest to softest power is on the order of a trillion.

A Network Orange

kind of average displacement the size of an atom. Not only does the human ear have such a wide dynamic range, but the brain somehow is able to allow simultaneous regions of the range curve to be activated, as exemplified in the cocktail party effect we discussed before. One way to summarize the exquisite design of the human ear is to say that it handles unpredictable data well.

Next, consider the human eye as an input peripheral. Dynamic range is also a specialty of the eye. In fact, there is a pupil—a hole—in the eye that expands or contracts, letting in more or less light to achieve extra dynamic range. We mentioned before the remarkable signal processing that enables us to perceive the full moon as a virtually perfect circle. The daytime sun is roughly a million times brighter than the full moon, yet people tend to guess the ratio to be in the thousands. The factor of a million is more believable if one imagines being in a completely dark room—allowing one's pupils to adjust to the darkness—and then suddenly rushing out into midday sunlight. Now *that* will be a bright light! Thus the eye, by virtue of its tremendous dynamic range (it is believed that at the sensitive end of this range, the eye can sense just a few light photons at detection threshold), its signal processing, and its mysterious interplay with brain processing, is a fine peripheral indeed.

We mentioned in the previous chapter that the ear/eye system can cotask on the problem of vibration detection at the subsonic threshold. Here is another example of cotasking: Humans can compensate for poor vision by echolocation—a kind of sonar that uses the sounds of footsteps, say, rebounding off walls. It is as though one peripheral takes over for another. This skill is evidently a way of negotiating a highly variable environment.

The great dynamic range of human peripherals is just an example of exquisite design, not the final word on efficient I/O. These peripherals are capable of handling a wide spectrum of circumstances, the spectrum of what we might call "chaotic data excursion." The principle here can be summarized thus: The (conscious) human benefits from exquisitely designed peripherals, and furthermore these peripherals can cotask, with each other and with the brain, with respect to chaotic data excursion. The operative words are *exquisite* and *chaotic*. What is especially intriguing is the possibility that what we call human consciousness evolved chiefly in response to the driving force of a highly variable environment. It is intriguing, for example, to contemplate not just the dilation of the eye's pupil but also the way this simple physiological action evolved together with night vision, color perception, retinal preprocessing, and the brain itself.

Apparently, the evolved human brain draws awesome—one is tempted to say unthinkable—processing advantage from its active peripherals.

OUTPUT MODES: EXPERT SYSTEMS AND INTELLIGENT AGENTS

In the last two decades, there have emerged certain computer systems whose operation and definition are central in the AI field: expert systems and intelligent agents. For a sense of the difference between these, imagine a terminal screen in a doctor's office. The ill patient who has called and made an appointment is asked to sit down and answer questions about symptoms. Is there fatigue? Are there chills? Now that is an expert system, and a good one indeed if it finally outputs a correct diagnosis such as "You have type-A, 1984-pig-strain influenza." On the other hand, say a patient walks by and the system speaks up: "You do not look well; you had better sit down and answer some questions." That is an intelligent agent. An intelligent agent is expected to be proactive, as though it had a mind of its own.[21]

Several features of expert systems and intelligent agents are relevant to the problem of machine consciousness. First, it would be difficult for anyone to deny that the medical-diagnostic expert system described above is a useful idea. Certainly, in cases where a massive symptom database is enough to obtain the correct diagnosis, a machine could be more efficient than a human physician. We authors see nothing wrong with machine diagnosis, as long as checks and balances exist to guard against machine error (conversely, an expert system can sometimes guard against human error). In such cases it is hard to deny the efficacy of this rudimentary form of AI. Note that the expert system is valuable because of its *output:* the final diagnosis. As for the intelligent agent that voices concern when an ill-looking patient appears, this much harder problem can be said to be unsolved. If a patient who appears pale or gaunt is flagged by such a system, what does the system do with someone wearing warpaint or special cosmetics? Even if an intelligent, proactive diagnostic machine could handle variable input in the form of patient appearance or faculties, it is hard to imagine such a machine emulating the classical, stethoscope-carrying human doctor across the full spectrum of patient inputs. An intelligent-agent machine doctor might be able to take temperature, but the com-

---

[21] "Agent Theories, Architectures, and Languages: A Survey," in Wooldridge and Jennings 1994.

plexity of integrating this reading with an overall impression of the patient is overwhelming.

The following passage illustrates why we have come to think of intelligence as predominantly an output phenomenon.

> [The eminent computer scientist] Douglas B. Lenat started his career in computer science with a program he originally called Automated Mathematician [AM]. . . . AM started with notions of mathematical sets, and operations like set union, intersection, and difference. It went on to derive addition, multiplication, primes, and exponents. It even rediscovered Goldbach's conjecture. . . . Although the program successfully reconstructed the basics of arithmetic and elementary number theory, Lenat saw it thrash around aimlessly as it struggled to discover more. Lenat decided that for a computer to discover new facts or ideas, it had to be primed with related facts and concepts. For the past decade, he has tried to encode the knowledge of modern-day North America into a computer program called Cyc (from encyclopedia). The largest scale AI experiment ever to be attempted, Cyc may lead the way to extremely intelligent agents or may flop. . . . There is always a whiff of "it will never work" when one talks to most computer scientists about AI, but why shouldn't it work? More than 4 billion human computers are good at it.[22]

This passage offers many clues to the epistemological issues at hand. Note the word *struggled,* a subtle anthropomorphism, when what is meant is that the machine did not produce enough satisfactory output. Note also the patently output-oriented character of the Cyc machinery: It seems that the input to this gargantuan intelligent agent is the simple phrase *North America,* yet we expect a torrent of facts and inferences to follow as output. It may be an oversimplification, but we can think of Cyc as potentially intelligent (on the subject at hand); yet for consciousness to appear, the input would have to be vastly more variable. To put it another way, the phrase *it [AM] had to be primed* is a telling one; the AM machinery has an inherent input weakness. Finally, the closing sentence illustrates a tenet that, like the authors quoted, we also insist on. We do not believe there is any fundamental reason why blockade to AI cannot resemble what humans do. Forget that the quote amounts to a tautology (AI is defined in circular fashion): the key obstacle to true machine consciousness is that machinery such as AM or Cyc is not of the basic type to reach such a goal. A great deal of current AI represents a line of research and development that, no matter how long it evolves, may never work.

---

[22] Shasha and Lazere, 1995, p. 207.

Thus the state of affairs at the end of the twentieth century is this: Expert systems can now produce useful output and in this way seem intelligent. Intelligent agents, which are thought of as advanced, proactive expert systems, also produce output but so far handle environmental input in rudimentary fashion. Many issues have arisen concerning what computer architecture to use for intelligent agency, the meaning of will, and so on.[23] With these issues in mind, we turn to the kind of task at which humans are still the reigning experts.

The Mysterious "Gedankenexperiment"

Moving away from discussion of actual peripheral design and the notion of expert systems, we consider here some specific processing modes in which humans still retain a marked advantage over machines.

The "Gedankenexperiment"[24] is the canonical example of how the human brain can react to unpredictable input. Imagine asking a machine—through its microphone peripheral, say—a question involving heretofore unstudied physical phenomena. The reason we say "unstudied" is that many historical open problems of science were solved by way of Gedankenexperiment.[25] Evidently, humans are capable of fabricating experiments, even in the absence of real materials, in order to imagine solutions to problems. The Gedankenexperiment is a manifestation of the human brain's ability to say, "Such-and-such an input stream is new to me. I do not have sufficient materials for solution of this surprising new problem, so let me make internal tools and proceed. . . ."

In the following amusing but instructive anecdote, one can see clearly how a human mind will perform quite differently than a modern machine, and will emerge as superior to the machine, despite a vast difference in memory size (in either direction) or in megaflops (always in the machine's favor). It is one of many legends at the California Institute of Technology

---

[23] Wooldridge and Jennings 1994.

[24] *Gedankenexperiment* means "thought experiment," performed as though setting up a mental laboratory and carrying an imaginary experiment to its natural, intuitive conclusion.

[25] The early twentieth century yielded some of the great examples of the Gedankenexperiment. Albert Einstein knew that $E = mc^2$ could be derived simply by considering a lamp emanating light energy (and therefore an equivalent mass) inside a closed train car. Niels Bohr, another master of the Gedankenexperiment, once shattered a certain construction of Einstein's (Einstein had been concocting devilish Gedankenexperiment attacks on the quantum theory of Bohr and others) by describing a Gedankenexperiment to which one of Einstein's very own celebrated theories was central! See Jammer 1989.

A Network Orange

that Richard Feynman, himself a master of the Gedankenexperiment,[26] was once asked how to solve the following problem: Should a swimmer shave all the hair off his or her legs in order to maximize speed during competition? Apparently, the swimming coach had actually tried having some swimmers shave their legs, but the lap time differentials were inconclusive—fractions of a second, say. And yet the question would be important to a coach, who wants to gain every tiny advantage. Consider first a machine solution to the question.

Machine solution:

> HAIR: = NO;
> /*Iterate complex hydrodynamical swimming model.*/
> HAIR: = YES;
> /*Iterate complex hydrodynamical swimming model.*/
> /*Report difference in swim times.*/

This machine program reads like a typical supercomputer application, and in some ways it is typical: It is easy to state but requires tremendous CPU power, not to mention a sufficiently deep hydrodynamical model. Feynman's solution, on the other hand, is a jewel of mental virtuosity.

Human Gedankenexperiment solution:

Shave just one leg of a swimmer and see if the swimmer swims in a circular arc.

Marvelous! Amusement aside, the example illustrates a central point: Both the supercomputer and the physicist have exactly the same inputs (speed is the parameter, leg hair is binary yes or no, and so on). But Feynman was trained in the matter of physical symmetry, and like all good physicists, he was capable of applying it at just the right point in the analysis. Of course one can claim that, armed with Feynman's swimmer Gedankenexperiment, one can go back to the supercomputer, analyze the swimmer with one shaved leg and indeed see whether such a swimmer will describe a circular arc. This might even give a realistic answer to the coach's original dilemma. But we do not know today how to coax Gedankenexperiments from the machines themselves.

---

[26] So effective was Feynman with the Gedankenexperiment that even referring to him as a master smacks of understatement. His contributions to the understanding of fundamental particles, and the Gedankenexperiment origins of such understanding, are legendary. One could reasonably contemplate an entire work based on the Gedankenexperiments of Feynman.

Where would Gedankenexperiments fit into the problematic world of AI? It would be beneficial to have a Gedankenexperiment solution to earthquake prediction—a computer program, say, that could tell us what to add to the known inputs (earthquake fault stress, fault history, magma displacement, and so on) in order to solve the problem of what parameter space is large enough. As it is, current seismological computer models are more or at less of the qualitative form of the machine swimmer program above: One guesses at a model and tries input data on that model.

Likewise, it would be good to have effective Gedankenexperiment software in general scenarios for which the parameter space is incomplete. One might envision for the future a Gedankenexperiment-based computer technology in which machines possess internal "laboratories" and generate internal data to augment the usual physical data. Here is an example that might not work out but that illustrates the possibility: Apply supercomputers to the problem of speech recognition via the modeling of ear-brain subsystems. Perhaps a massive search over a galaxy of ear-brain neural networks would be more fruitful than the prevailing approach of choosing a particular ear model, being saddled with a frozen brain model (the computer used), and then training the resulting rigid system. If only we could understand the mysterious Gedankenexperiment of which humans are capable, we might make inroads into AI research.

A THEORY OF MACHINE CONSCIOUSNESS

Having scratched the surface of AI failures, noted the drawbacks in typical machine approaches, and acknowledged some progress in AI, we enter the realm of frank speculation. We believe it is better to conjecture in a virtually unrestricted, untested way than to relegate machine consciousness to the limbo of hard problems. The key to our speculative construction is the natural asymmetry between input and output functions.

Anyone who has worked with computer peripheral design knows that regardless of all the symmetrical textbook block diagrams and simplified computer-fundamental overviews, input and output functions are fundamentally not equivalent. This is a hard pragmatic fact, and it furthermore is demonstrable on fundamental scientific principles. Their basic asymmetry follows from simple thermodynamics: The entropy of inputs is generally higher than the entropy of outputs. This is true for the almost trivial reason that inputs are by nature unknown, whereas outputs are system-determined. Engineers know that "input software," or software drivers (for microphones, cameras, and external devices in general), are usually more difficult to program than output drivers (to speakers, video displays, and so on). An input driver software designer has to anticipate what the future holds; an output design involves always-known computer data, and outputs can be driven at the time of the computer's choosing. By contrast, an input subsystem must deal with the demands of a usually whimsical environment.

To render this essential notion more clearly, we state

> **Proposition 1:** *Input and output phenomena are genuinely distinct, the asymmetry being that input is often possessed of chaotic features, not to mention high entropy, and is far more problematic.*

Our approach to a theory of machine consciousness, then, starts by adopting the descriptors *intelligence* and *consciousness* as primitives and augmenting the first proposition with another.

> **Proposition 2:** *Intelligence may require sophisticated input processing but is manifest chiefly by effective output behavior, whereas consciousness always involves effective input processing.*

We have inherited from Weizenbaum[27] a fine instance of human perception of a particular machine as intelligent. In Weizenbaum's ELIZA example, people believed they perceived intelligence in the machine's responses. We are not saying that ELIZA software was intelligent, and in fact we empathize with Weizenbaum's horror at the ease with which human users fell into this belief. ELIZA cannot be "intelligent" according to Proposition 2, as long as we think the output behavior of ELIZA was not effective *enough*. Though almost none of the machines and software in the history of computing seem to us intelligent, we submit that when machine intelligence really does systematically emerge, the key will be effective output functions.[28]

> **Definition 1:** *Machine intelligence is the ability to produce a manifestly relevant output stream.*

---

[27] Weizenbaum 1976.

[28] A simple example of effective output is the emanation of grandmaster-level chess play from the IBM machine "Deep Blue." It could be said that Deep Blue produces the kind of output necessary to win the "intelligence" descriptor (after all, *something* should be attributed to a machine that can beat any or virtually any human), but as we stressed before, Deep Blue does not really play chess in the full spirit of chess, and within our current theoretical framework, it certainly is not conscious.

Definition 2: *Machine consciousness will be a state of legitimate and effective anticipation with respect to chaotic and entropic input streams.*

We can paraphrase the definition of consciousness as follows: "The conscious machine will expect and be able to process meandering input data." Consider now the brilliant ideas, Gedankenexperiments, and theoretical constructions of Norbert Wiener, who wrote,

> Let us consider the activity of the little figures which dance on the top of a music box. They move in accordance with a pattern, but it is a pattern which is set in advance, and in which the past activity of the figures has practically nothing to do with the pattern of their future activity. The probability that they will diverge from this pattern is nil. There is a message, indeed; but it goes from the machinery of the music box to the figures, and stops there. The figures themselves have no trace of communication with the outer world. . . . They are blind, deaf, and dumb, and cannot vary their activity in the least from the conventionalized pattern.[29]

Wiener's prophetic words (four decades old!) cut to the heart of the matter: Communication with the outer world is essential for emergence from the limbo of mechanism. Yet another classic Wiener example of "intelligence," which again fits our input/output paradigm, is the spectacular example of the mongoose and the cobra. As they confront each other in their deadly dance, the mongoose joins the cobra in a sideways oscillation. But the mongoose, presumably drawing on its "more nonlinear" nervous system, manages to advance the swaying phase a little on each pass, until it is out of phase with the cobra. The cobra evidently does not react well to this phase shift; it keeps to a steady swaying frequency. In time, the mongoose manages to attack the cobra from the side—180 degrees out of phase—and kill it. We might say the mongoose is more intelligent, by virtue of its superior output behavior.

Wiener's "music box" and "mongoose" scenarios lead to a seemingly inescapable conclusion: No matter how sleek a system's behavior, one must always ask about its ability to gather and process I/O. No one will deny that the cobra is efficient, undoubtedly having evolved to a sharp local optimum in some appropriate biological fitness parameter space. But the mongoose evidently exploits the cobra's limited I/O-processing capability. We are not saying the cobra has dulled input peripherals—some of its senses are sharp as can be—but only that its processing of input is relatively dull, or dull enough to be fatal. Remembering also the incommunicative, deter-

---

[29] Wiener 1954, pp. 21–22.

ministic figurines of the music box, we shall say that though intelligence may be manifest, the property we call consciousness may not be present. Imagine a music box sublimely intricate and supercomputer-driven, a whole city of figurines carrying on conversations; perhaps the whole mechanism even exhibits some environmental sensitivity but is otherwise largely deterministic. Although intelligence might be evident, without sufficient input processing we cannot consider this intricate mechanism conscious.

Thus our highly speculative thinking and our intuition lead us, on the basis of our propositions and definitions, to the following claim.

> **Conclusion:** *Computers do not yet possess consciousness. They exhibit intelligence only in isolated circumstances. For consciousness to emerge, there will have to be a revolution in the matter of input processing.*

Though our ideas are conjectural and unproven, there is historical precedent at least for the fundamental notion of the dependence of human consciousness on the ability to process input. Such ideas are in fact centuries old. Consider this observation by Dreyfus:

> . . . although not aware of the difference between a situation and a physical state, Descartes already saw that the mind can cope with an indefinite number of situations, whereas a machine has only a limited set of states and so will eventually reveal itself by its failure to respond appropriately. This intrinsic limitation of mechanism, Descartes claims, shows the necessity of presupposing an *immaterial soul.*[30]

Dreyfus goes on to point out that because a computer of today has an overwhelming number of possible states,[31] "it is not clear just how much Descartes' objection proves." Along similar lines, it is well known that we now have machines with more memory cells than a human brain possesses. Hence it is natural to ask why there cannot exist a "Descartes machine" of modern design that would *not* "reveal itself" as an impostor. Consider this paraphrase of a portion of the foregoing quotation:

> *that the* mind system *can cope with an indefinite number of* chaotic scenarios, *whereas a machine has only a limited* number of degrees of freedom with respect to I/O. . . .

From this simplified, modernized, and transformed perspective, then, we do embrace a Cartesian theory of consciousness. We replace the lack of

---

[30] Dreyfus 1993.

[31] Dreyfus gives an estimate of $10^{10000000000}$, an unthinkably large number of possible internal machine states, though still not as large as the number of possible games of chess.

system states with an inability to achieve a "legitimate state of anticipation" with respect to chaotic streams. We do not say a manufactured machine cannot achieve this legitimate state, only that no machine has yet done so. Conversely, it may be that humans evolved what we call the conscious state through little more than long-term selection pressure to negotiate a highly variable environment. It will not be possible to achieve machine consciousness until sufficiently rigorous attention—might we even say homage?—is paid to the chaotic-stream, high-entropy anticipatory processing that human consciousness exhibits.

References

Dreyfus, H. L. 1993. *What Computers* Still *Can't Do: A Critique of Artificial Reason.* Cambridge, MA: M.I.T. Press.

Frey, P. 1983. *Chess Skill in Man and Machine.* New York: Springer-Verlag.

Jammer, M. 1989. *The Conceptual Development of Quantum Mechanics.* Los Angeles: Tomash Publishers.

Junqua, J.-C., Mak, B., and Reaves, B. 1994. *IEEE Trans. Speech Audio Process* 2(3), pp. 406–412.

Mellish, C., ed. 1995. IJCAI-95, Volumes I and II of *Proceedings of the Fourteenth International Joint Conference on Artificial Intelligence.* Montreal: Morgan Kauffman Publishers.

Newborn, M. 1996. *Kasparov Versus Deep Blue.* New York: Springer-Verlag.

Shasha, D., and Lazere, C. 1995. *Out of Their Minds: The Lives and Discoveries of 15 Great Computer Scientists.* New York: Copernicus/Springer-Verlag.

Weizenbaum, J. 1976. *Computer Power and Human Reason.* New York: Freeman.

Wiener, N. 1954. *The Human Use of Human Beings.* New York: Da Capo Press.

Wooldridge, M., and Jennings, N., eds. 1994. *Intelligent Agents.* Berlin: Springer-Verlag.

MULTIMEDIA:
MÉLANGE OBSCUR

There is now a great deal of ongoing work on multimedia systems . . . within the database community, as well as outside it. All of these works, without exception, deal with integration of *specific types of media data;* for example, there are systems that integrate certain compressed video-representation schemes with other compressed audio-representation schemes. However, to date there seems [to be] no unifying framework for integrating multimedia data which is independent of both the specific medium and its storage.[1]

A NIGHT AT THE OPERA

Imagine a night at the opera, a first-rate opera. Costumes and stage decoration are expected to be in keeping with the operatic theme and subthemes and with the demographic and historical import of the opera. The instrumental music is consistent with the vocal passages. The vocal music, in turn, expresses the composer's thematic intent. You hold in your hands a program elegantly scribed on fine paper, a summary description of the event. The real-world operatic multimedia—visuals, sound, documentation, and even

---

[1] S. Marcus, "Querying Multimedia Databases in SQL," p. 276 in Subrahmanian and Jajodia 1996.

43

such abstract values as drama, history, and composition—cohere in a manner we have come to appreciate as cultured, refined. The effectively presented opera is a superb example of harmonious media mixing—in literary terms, a *mélange de genres*.[2]

Now consider an amateur production rife with inconsistencies. The instrumental music is too loud or too soft in relation to the vocals. The written program you hold is of notably inferior quality. One might call this unfortunate opera a *mélange obscur*, a phrase we hereby coin to mean an incoherent mix, a slipshod construction, an ungainly design.[3] One envisions cheerleader outfits worn in an Italian opera, with background stage sets representing the Great Wall of China. Even more distracting is the juxtaposition of various widely varying quality levels. Perhaps you hold a program set in Cyrillic font, the ink bleeding into the cheap paper. The event would probably be painfully funny; at best, in the words of Steinitz, it would fill the astute observer "with artistic horror."[4]

Many modern-day instances of computer-based multimedia amount to a *mélange obscur*. Too often one scheme, developed by company A for sound, is mixed with B's scheme for images, hypertext integration software from C, and so on. There is no standardized multimedia structure, and even more important, no consistent standard of quality has been imposed. Let us explore some examples of the *mélange obscur*. You open a hypertext encyclopedia and want to learn, see, and hear about the Siberian tiger. It may well happen that you get a mottled picture of the tiger, because a low-grade compressor/decompressor has been invoked for fast image rendering. The sound describing the tiger is likely to be of low fidelity. Perhaps the original narrative—together with natural jungle sounds—was mixed as stereo, and you are getting only the left channel because the resident digital audio software supports only monaural playback. The text description on the screen may involve too many fonts, creating a garish effect overall. The typeface quality may be much better or much worse than the image quality. The disparity between levels of qual-

---

[2] This wonderful phrase has been used in reference to the well-coordinated mix of genres found in the works of Proust.

[3] We intend this phrase as antithetical to the aforementioned *mélange de genres*. We thus reverse the notion of a successful mix by enlisting a phrase employed by Mallarmé in his poem "Le Tombeau d'Edgar Poë." It was widely believed in Mallarmé's time that Poe destroyed himself by imbibing *mélange obscur*: badly blended wine.

[4] The great chess master once said that an unsound sacrificial combination, no matter how showy, would fill him thus.

ity across types of media in itself undermines the aesthetic effect. (In some ways, this is worse than having every type be of equivalent low quality, which at least offers consistency.) Then there can be real-time problems. For example, the tiger image may be the key frame of an actual tiger video. One starts the video and the tiger moves. But does it move in synchrony with a soundtrack? Maybe not, depending on computer platform, processing power, and so on. And the motion picture may be of even lower quality that the still frame.

Why do we freely tolerate the *mélange obscur* in computerized multimedia? One answer is that, for poorly understood reasons of technological culture, people are quite patient not just with multimedia but, alas, with computerized *anything*. Perhaps this is because something computerized is something not yet grown up, something unformed, like a child who is forgiven for not adopting the adult perspective.[5]

But another answer to the question of why we tolerate the current woeful state of multimedia is that for various reasons, it is about the best it can be. This in turn is a consequence of an abundance of technical problems, especially real-time problems of bandwidth and memory capacity. But before these technical problems can be successfully addressed, and before the cultural expectation of coherent media can be raised, the bottom-line problem of multimedia unification must be solved. Therefore, we shall concentrate for much of this chapter on the issue of unification.

As Marcus forcefully points out,[6] little work has been done in laying the mathematical foundations of multimedia technology. It is important here to realize two things. First, even if there were a unified foundation, many users would not care or notice, and that is fine. The final result of a successful unification of multimedia should be a *mélange de genres*—a harmonious, coherent product. Second, the opacity of mathematical unification to end users would not make such mathematics any less essential. Most computer users are not also mathematicians, but whether they play games, process text, or do taxes, a strict mathematical foundation underlies

---

[5] Along these lines, one of the authors recently had a lively discussion with a computer scientist on the observation that students in the field do not expect to learn the *history* of computer science, although everyone understands the importance of the history of physics, biology, and chemistry. It is as though computer science itself is deemed so fresh and new that students feel they are riding the leading wave of a "revolution," when in fact, there have arguably been three or four computer science revolutions since World War II.

[6] Marcus, in Subrahmanian and Jajodia, 1996, p. 263.

the very CPU operations of the machine and the way in which lines of software code work together.

Whatever the reasons, modern-day multimedia applications are fraught with mismatch and inefficiency at the user level. For decades there has been a notable paucity of standards.[7] Our purpose here is first to define and outline the components of modern multimedia and then to indicate how the various pieces might be unified. Finally, beyond the unification issue, we shall sketch a kind of "Frankenstein effect": how multimedia might evolve into a force that overhauls the very field that gave birth to computers and computerized multimedia in the first place—science itself. The reason we expect the practice of science to sustain unprecedented influence from multimedia is as follows. The design and operation of most classes of multimedia-handling computers began with graphical representations of simple geometrical objects. The next phase was visualization, involving more complex shapes with which to reveal relationships among data. From the 1960s to the present, scientific visualization has grown into a massive industry and, perforce, into a major multimedia sector. We shall be claiming that the story does not end here and that science—more precisely the way we do science—has much more to fear in the future.

THE MEANING OF MEDIA

The etymology of the word *media* runs deep.

The original Latin *medium* meant "middle" or "mean." A person can be of medium height or weight. But a fish subsists in a fluid medium; the animals of the world flourish in different media. What brings these meanings together intuitively is a word such as *midst:* An entity may exist at the midst (or middle) of some scale or in the midst of some environment. In usage related to the idea of a medium as a substance, an artisan may choose to work in one of several media (liquid, powder, clay, plastic, or whatever). On the notion of a surrounding fluid, it was not too long after the invention of television that one could speak of TV as a medium.[8] The

---

[7] In the 1990s, roughly the fifty-year mark for the history of computing, some multimedia standards are gelling. Unfortunately—and the details lie beyond the scope of the present book—there is a complex relationship between standardization and commercial monopoly. It may turn out that to achieve effective multimedia mixes, monopolies need to be avoided or blunted.

[8] Actually it was in the 1920s, before television, that the notion of *medium* as a surrounding "information fluid" began to be used. One can think of early radio and newspapers as media.

concept of television carries with it both the notion of a surrounding (indeed a saturating) fluid and that of an information "channel." A spiritual medium supposedly acts as a conduit between the living and the dead. We can think of computerized media, then, as conduits of information transfer. At the inception of television, it may have been natural to think of a glass screen as a canvas, with electronic "paint" as the medium. But in light of today's reliance on information density and the explosion of graphical visualization, it is more natural to think of the television mode itself as the medium, the visual display being just a user's terminus of the action. Thus we take something like a "multimedia document" to be a user's terminus, at which various conduits such as visual, audio, maybe even tactile media act.

When dealing with multimedia data, one is dealing not only with the data but also with an *intent*. We think it is important not to become overly obsessed with the user's terminus of a medium, or even with the media data; rather, one must consider the entire process, from intent to presentation. With this in mind, we turn to discussions of various multimedia data types per se, then later to the problem of unification of types.

Visual Data

There is good news and bad news about the current state of visual data processing (essentially, images and movies) within the multimedia context. Two parcels of good news:

1. Visual data are often highly compressible, so that an original image can be compressed as much as 25:1 without apparent degradation, and a motion picture can be compressed as much as 100:1.[9]

2. Color displays are quite prevalent in the 1990s; color is no longer a rare luxury.

The bad news:

3. Visual data tend to be voluminous prior to compression.

---

[9] These approximate values are chosen to convey a sense of scale. Images starting at 24 bits per pixel (typical red-green-blue color format) can be compressed to about 1 bit per pixel (via, say, the JPEG compression algorithm) with only slight degradation. With movies, the MPEG algorithm goes further by taking motion redundancy into account. For very high-quality compression these ratios are too optimistic, but they are adequate guidelines for "good not excellent quality" storage.

4. Color displays are not realistic. For example, a computer display's dynamic range limitation is severe compared to the corresponding range of the human eye.

5. Typical displays must render any three-dimensional object in two-dimensional mode.

On bad news item 4, think of a real-world sunset, the sun low in the sky, yellow but still very bright, the sky varying from orange near the sun to deep blue way above. Many modern CRT displays can do pretty good versions of the orange sunset color and the deep blue sky above. But they fail miserably in representing the bright sun. Until just before sunset, the setting sun is really too bright to gaze on. A display screen, however, can get only so bright, certainly nowhere near bright enough to model any but the final stage of sunset. The dynamic range of the human eye is about 10 billion, the approximate ratio of the brilliance of the brightest sun one can stand to that of the dimmest evening star one can see. But the dynamic range of a typical computer display is something like 300:1, or, if one wants to spend a great deal of money, perhaps a few thousand to one.

On bad news item 5, one might argue that even in the real world, we see a two-dimensional image; after all, the retina of the human eye is essentially two-dimensional. But in the real world a slight motion of the head reveals perspective. Also, we have two eyes to yield three-dimensional information via parallax. It is true that modern laboratory experiments in virtual reality address some of these issues, but by and large, the common computer system has a two-dimensional display.

Tremendous effort has gone into the generation of two-dimensional images to make them appear three-dimensional. Surface rendering, shading, and so on add reality to a display. A worthwhile goal for the future would be a holographic display system for the common citizen: a three-dimensional display one could literally "walk around."[10] It is already possible to fabricate actual, physical curved surfaces via computer control. What we need is a reduction of several orders of magnitude in fabrication time and a corresponding boost in the degrees of fabrication freedom. The world of machine multimedia will have really gotten somewhere when

---

[10] There is nothing preventing holographic displays in principle. The technical problems abound, however. For one thing, because of the natural wavelength of visible light, more than 10,000 pixels per inch would be needed (somewhere in the system) to allow true holography. This is something like 100 times the linear density of a modern display screen, which means 10,000 times as many pixels on a screen, giving rise to severe memory and bandwidth problems.

one can program up new furniture or a flimsy but stable kite to fly (with an accompanying ball of string, of course).

AUDIO DATA

Audio data storage per se is not so problematic as video storage. It is interesting, however, that although the human ear is more forgiving than the eye in one sense (sound requires much less bandwidth), the ear is more demanding in a second sense (it is very hard to compress sound and maintain subjective fidelity). Thus the primary technical challenge of sound media handling is to maintain sound fidelity without adding too much to the visual data size. When visual data are highly compressed, high-fidelity sound must not be allowed to consume a comparably high portion of the bandwidth.

Though storage is not a primary stumbling block for sound, fidelity is a problem. Yet nearly all multimedia systems are limited by poor sound output devices, such as miniature speakers or buzzing or humming from sound cards placed inside electrically noisy personal computers. It is as though sound capability were a retrofit (add-on) feature, and in fact, it almost always is. The same can be said of the tiny lapel-type microphones used for computer sound input. Sound I/O is not yet completely integrated into the computer world. Yet such integration is essential if multimedia products are to attain a respectable level of excellence.

TEXT STILL SUFFERS

The basic story of the evolution of computer-displayed text is not well known. For two decades after the 1950s, the primary mode of text display was via alphanumerical circuits and chips.[11] This means that computers stored ASCII codes, and when the visual display electronics encountered a certain number (say 65 decimal) in memory, a certain character (say "A") was drawn on the screen. And *this* means that only an extremely limited character set—just a few dozen characters—was possible. Even today, some so-called dumb terminals still employ this scheme of text display.

---

[11] We are concerned, as usual, about large-scale trends—in this case, the issue of when the alphanumerical displays dominated in the general population of users. The distinction is important; for example, the military "Whirlwind Project" culminated in the early 1950s with an oscillographic display that could effect line drawings such as graphs of trajectories. But this was a very large, power-hungry, and idiosyncratic piece of special hardware (Augarten 1996).

But in the early 1980s there emerged, on a large scale, displays in which every character was to be drawn out of pixels. Now we can have arbitrary fonts, crazy characters, and foreign alphabets just by changing the set of pixels to be drawn for each of the (many more than a few dozen) character codes. Later, another leap in generalization took place: Scribing languages emerged.[12] With such a language, one's computer can actually store not just crude, letter-by-letter codes but also a formal description of the text to be drawn, including the font(s), size(s), attributes (bold/italic), and so on. An extra advantage of scribing language, beyond its obvious flexibility, is its independence from resolution. A program can be written that makes relatively ragged characters when used with a low-resolution visual display, but when the same program is run on a professional printing device, the print can appear flawless.

So isn't text, within the multimedia context, doing well by way of this positive evolution? Yes and no. Yes, because the flexibility and high quality of modern text-display schemes are impressive and good for multimedia in general. But no, because there are perpetual performance problems. A truly advanced text processor, especially when it is part of a larger multimedia package, requires something approaching supercomputer power to render all text operations (even just the display functions) seamlessly in time. Anyone who has experienced a high-quality text display has seen how the fundamentally simple idea of scribing text to a screen causes even the fastest widely owned CPUs to falter. The basic reason for this performance problem is that given the evolution of pixel-oriented character display, machines have to draw, in some sense or another, every single pixel according to software specification. It is true that whole sets of pixels, such as the most commonly used whole characters of pixels, can simply be copied from memory to screen quickly, but when the encumbered CPU encounters anything "strange," such as on odd symbolic character (as in mathematical typesetting), or a line drawing (which is entirely unpredictable), or a surprise font change, pixels must be drawn according to some mathematical scheme, and that is very expensive.

In spite of tremendous advances in the handling of the text medium, the power of modern machines is out of sync with the demands of arbitrary text display. The situation was actually much worse in the 1980s, when pixel-by-pixel text display emerged, for the CPUs were much slower

---

[12] One such language is PostScript, with which text, graphics, and so on can be mixed in quite arbitrary but mathematically defined ways.

then. Indeed, one characterization of the elegant new displays of that decade was that they were "slow." We believe that long-time users are especially patient now with today's time-domain stuttering and CPU pauses because of the stark improvement over what went before. (This is yet another example of the user forgiving computers for not having completely grown up.)

G. Davenport offers a sharper critique of modern text media and of the useful operations that one should be able to perform on them.

> Is there any sign of intelligent life beyond keyword searching? My word processor's spellchecker constantly urges me to substitute "composting" for "compositing." My Web search for the proverbial "Genie in a bottle" instead yields "Gen-X in a battle" with a 12 per cent likelihood of correctness. Should I consider this brilliant, witty, and insightful commentary from my user-friendly laptop computer? Or is it just plain stupid? Oh, brave new world, that has such software in it![13]

Davenport goes on to point out the talents and qualities of a "favorite research librarian."

> Her ability to interpret my inquiries, knowledgeably expand them, and then extrapolate them into a rich model for search and retrieval makes her an invaluable and pleasurable resource. She takes pride in knowing her library thoroughly, both spatially (where to find a book) and temporally (how her book inventory has changed and evolved). As she formulates her plan of attack, her sophisticated understanding of language, culture, experience, and other knowledge-domain models simultaneously converge and are re-mapped onto the reality of her library.

On the limitations of current multimedia-based "storytelling," Davenport again notes that we pretend the multimedia machinery is capable of human mimicry (even if we did not pretend, we might liberally forgive the machinery for its shortcomings). In Davenport's words, "We appear hell-bent to drag simple models into the realm of storytelling and make them the foundation of automated storytelling systems."

In some sense standing at the geometric mean between text and planar drawing on the one hand and images and movies on the other, is the problem of three-dimensional rendering of system objects. Everyone knows that in modern times, the pixel-by-pixel character motif made possible the drawing of icons, those small drawings that represent files, programs, or the like. Later it became possible, by using gray scale, to render

---

[13] Davenport 1996, p. 10.

icons somewhat three-dimensional in appearance.[14] Then, of course, as machines got still faster, it became possible to have full-color drawn objects on a citizen's display screen. And matters will not by any means stop here. The phenomenon whereby the demands of the text medium "pull machine performance along" is likely to take the form, over the next two decades, of three-dimensional *rendered* objects. It is true that one can now invoke sophisticated software with which to render (usually slowly), say, a yellow-plaster, impressively illuminated bust of Napoleon. But one cannot yet get this kind of functionality with all the attendant movability, instantaneous drawing, and so on that are to be found in an adequate icon-based display.

We summarize the coherence problem with the text medium as follows: Text and drawing can be said to be more advanced than the other media, which is just one manifestation of *mélange obscur;* yet through the decades, the computing machinery finds itself in a perpetual phase lag.

INK AS DATA MEDIUM

Here at the threshold of a new century, there seems to be an ongoing text-media crisis. Generally speaking, text-processing software is not entirely resonant with human desire or skill. There are various manifestations of this crisis, the sad thing being that not a single one of them is in principle beyond repair.

An exaggerated but telling description of one aspect of the text-media crisis is provided by T.K. Landauer:

> Electronic documents would have long since arrived if it were not for one stumbling block: people hate them. In almost every trial of providing electronic documentation, users have found the documents so unpleasant to use and so unsupportive of the work that they are needed for that they have refused to surrender their paper. The result has been that where electronic documents have been introduced, the paper documents that they were supposed to replace have remained. Thus instead of doing away with expensive paper and substituting cheap electrons, the organization ends up providing both.[15]

---

[14] Indeed, as soon as it became feasible to draw with black or white pixels only, ensuing generations of systems drew with "2-bit" gray scale, say, which would be black-dark-light-white, or four gray values. The drawing was correspondingly slower, but now the relevant icons, bars, knobs, buttons, and objects in general could be given shading and apparent depth.

[15] Landauer 1995, p. 249.

We concur that some people hate electronic documents, though the hatred seems confined to certain groups of workers and is not so universal as Landauer maintains. But let us agree that the hatred is widespread. Beyond the psychological issue, Landauer makes the interesting point that business documents are harder to read than they used to be, ostensibly because of the tendency for users to invoke right-justification in machine text processing.[16] This is a good example of computer software not being entirely in synchrony with human skills. Text justification, on most systems even today, is crude compared to the professional typesetting of a fine book. In principle, computers could easily be made to close this gap between human skill and software design, but this does not happen.

Another manifestation of the crisis involves degrees of physical freedom. A typical modern computer screen, though much more readable than the screens of a decade ago, is still woefully inadequate compared to having an open book or several books spread out on a conventional desk. Various modern operating systems endeavor to construct for the user a digital "desktop," but the computer media display is physically cramped relative to true desk work.

One positive development in the text-media field has been the emergence of hypertext.[17] It is generally believed that hypertext is an efficient way (at last) for computer users of today to traverse linked databases. But some issues are best resolved by experiment. The work of McKnight et al.[18] indicates that a document created *expressly* for hypertext software can still be easier to use—if the reader is resolving a given set of questions—when printed out on paper. In fact, students who took part in the experiment not only performed better when using the paper version but also indicated a preference for the paper. Of course, this experiment involved linked text, whereas a truly multimedia hypertext piece, involving, say, sound and color images, does not even *have* a paper counterpart. These

---

[16] Landauer cites, for example, a study claiming that it takes 10 percent longer to read a right-justified page than a ragged-right page, presumably because the jagged edge allows more rapid resolution of where one's eye currently is trained on the page, as that eye moves to locate the next line.

[17] Hypertext is a text-like structure in which "links" can be used to traverse vast, connected realms of information. One might, within plain text, for example, see visual links to large color images, and one could decide whether to follow such links. In this way, hypertext has the welcome feature of letting the reader (or maybe we should say the peruser) make key search decisions.

[18] "A Comparison of Linear and Hypertext Formats in Information Retrieval" in Macaleese and Green 1991.

observations pave the way for our proposition: The advantage of hypertext lies in the stark fact of multimedia. There is not necessarily any advantage in information navigation per se.

We believe that future designers of hypertext software, and especially hypertext documents, should pay attention to the physical and visual constraints compared with the relatively free medium of paper print. Unfortunately, it has been too easy for software designers to impress users with all the other freedoms: color combinations, sound options, motion picture options. Users in turn should not confuse these media freedoms with more rudimentary visual and physical freedoms. And above all, users should not automatically assume that the wonders of multimedia are replacing the conventional information modes of paper print.

Connected with the issue of paper print versus hypertext is the issue of typing versus drawing and handwriting. Why does virtually every modern computer user toil at a keyboard?[19] There is no telling what subtle forces account for the vast majority of work being performed via typing.[20] There are other options, such as the interesting idea of allowing hand-laid ink to be a genuine multimedia medium. Aref et al.[21] summarized the advantages of pen-based computing as follows:

1. *As notepad computers continue to shrink and battery and screen technology improves, the keyboard becomes the limiting factor for miniaturization. . . .*

2. *The pen is language-independent—equally accessible to users of Kanji, Cyrillic, or Latin alphabets.*

3. *A large fraction of the adult population grew up without learning how to type. . . . However, everyone is familiar with a pen.*

4. *Keyboards are optimized for text entry. Pens naturally support . . . a much richer domain of possible inputs.*

These are beautiful reasons because they are so humane. For example, reason 2 might just as well read "All literate peoples have the same basic

---

[19] There are exceptional cases, of course, such as computers that allow handwritten input. But it is fair to say that such devices have not yet "caught on."

[20] This notion of a writing technology overwhelming an older but better one is exemplified in the unfortunate experience of anyone who tries to use a modern "whiteboard" with dysfunctional, dried-up color markers. Even if the markers are wet and functional, anyone who has used old-fashioned chalkboards knows that chalk has much more dynamic range: One can make thin or thick chalk lines, but the wet-marker lines are basically frozen in width.

[21] In Subrahmanian and Jajodia 1996, p. 113.

hand coordination." The authors go on to describe how ink, *as ink,* should be considered a valid medium. Of course, the "ink" is to be stored digitally, but no short cuts to the storage are taken in their data model. One could even have the best of both worlds: A stored pen work might include regions that needed to be cast in digital ASCII form, but otherwise a signature or sketch would be stored *as is* in the relevant multimedia database. Certain sketches (such as an approximate circle) could be adjusted at the user's will into more accurate, digitally defined shapes. Some such automatic adjustments are possible now, with certain computers and software. But what is needed is a unified approach to ink media. The important gain is the extra degree of freedom afforded the user. Such a strong component must necessarily be easy for the user to invoke at will. Ideally, using ink media should be as easy and unfettered as conventional paper correspondence, but the user should also be able to benefit from computer-specific advantages: ink database queries, pattern recognition, and so on.

TELECONFERENCING AS CANONICAL TESTBED

Teleconferencing is the act of communicating, in real time, via multimedia.[22] This mode of communication—especially because of the real-time constraint—is the harshest and therefore most valuable test of multimedia schemes. A simple calculation shows just how difficult teleconferencing technology can be. Say we demand the option of a high-resolution transmission/reception of not just two participants but of groups of people at each end of the conference. There might also be reason to show drawings or other artifacts over the visual link. Thus the real-time "movie" to be transmitted might be 10 frames per second (not the best animation, but watchable) at a resolution of 640 by 480 pixels (roughly the resolution necessary to hold arbitrary figures up to the camera). If there is no video compression—that is, if every pixel is transmitted and received as is— then the bandwidth requirement is about 10 megabytes per second. Forget telephone lines; not even a typical, on-site network can handle this kind of bandwidth. And we speak of only one direction; the other party needs the same bandwidth!

---

[22] The typical realization here is the "picturephone," by which you would talk to and gaze at the other party, who in turn would be talking to and gazing at you. An actual business teleconference might involve two boardrooms, separated by thousands of miles, with the two sets of directors staging a joint audiovisual meeting. The notion of "real time," meaning multimedia reception changes in synchrony with real events, is really the hardest part of the technology.

Thus the multimedia problem for teleconferencing is severe, primarily because of video requirements. This is why many teleconferencing designs focus on spatial compression, reduced frame rate, and temporal compression (attempting to ignore image parts that do not move). A second problem has to do with audio, and this is not strictly a bandwidth problem. It is the latency problem: One cannot allow "breaks" in the sound, because the two teleconferencing parties find them too obnoxious. If the conference is taking place over a network and the network gets very busy, one can imagine sound being interrupted for a split second. If this interruption happens in the middle of a word, or even between words of a sentence, it is a serious defect. One option is to "buffer" the audio medium so that all sounds are, say, two seconds late and any "breaks" are absorbed by the computing machinery. But the effect is like that people experience talking between earth and moon: There is a horrid two-second delay. This latency problem is just one reason why network teleconferencing is not more prevalent. The ear is less forgiving than the eye when it comes to latency. (This is also true of slow-motion video; the audio equivalent of slow motion has little use.) A split-second interruption in visual data is simply not so awkward as the analogous interruption of speech.

As the most stringent form of multimedia processing, teleconferencing should stand as the canonical testbed of the future. We shall know that progress in multimedia unification has been made when teleconferencing is cost-effective, glitch-free, and easy to use, and furthermore mixes well with other multimedia formats.

A Scenario for Unified Multimedia

What is to be done about the *mélange obscur?* The bad news about current multimedia is the lack of standardization and the often cumbersome mix of separate technologies. Teleconferencing is an example of the cumbersome mix: Though custom systems are in use today, it is essentially impossible to add teleconferencing effectively to existing computer systems. If one obtains a device, for example, to add onto a personal computer so that images over a network can be received, such a device will probably stagger and pause when "something else" happens, such as the arrival of new e-mail or some other sudden demand.

Aside from the real-time limitations of bandwidth and latency, there exist problems with compatibility across products and systems. Most of these compatibility problems will never be solved unless a solid, unified multimedia framework is developed and adopted. To convey the difficulty of integrating visual, audio, and other database information in a multimedia presentation, Marcus[23] gives a rigorous example of how to quantify the qualitative query

"Is there an audio representation as well as a picture of a midsized car?"

as the formal logic query

($S1, S2, S3, C) frametype(S1, audio) & frametype(S2, video) & C e
flist(S1) & C e flist(S2) & type(C, midsize, S3))

and indicates further how this can be cast in SQL (query language).

```
SELECT L_1, Statename
FROM Frametype t_1, Frametype t_2, Frametype t_3,
    Type L L_1, L L_2,
WHERE Media Type = 'video' and
    t_2.Type = 'audio' and
    L_1.Statename = t_1.Statename and
    L_2.Statename = t_2.Statename and
    L_1.Featurename = L_2.Featurename = t_3.model and
    t_3.size = 'midsize'
```

We have displayed the three equivalent queries in the forms of English, formal logic, and query language to suggest how difficult it is to computerize multimedia. The simple English query sentence has turned into a paragraph of SQL code. What is worse, there are natural queries that cannot be programmed in the SQL fashion and so require a more general query structure of which SQL would be just one layer!

Difficult as the problem of multimedia unification may be, we applaud efforts to sound that problem's depths. As might be expected, any serious attempt to provide a universal multimedia framework is frightfully complicated, because it reflects the vast world of possible human activity, perception, and meaningful input states.[24]

---

[23] Marcus, in Subrahmanian and Jajodia, 1996, p. 272.

[24] "Towards a Theory of Multimedia Database Systems," in Subrahmanian and Jajodia 1996, contains an interesting but very complicated proposal for a completely general multimedia processing language. It is possible that, complicated as their model is, one cannot do better.

Scientific Visualization and the Demolition of Science

Scientific visualization is one of the primary forms of multimedia presentation in both commercial and educational sectors. There is no doubt that scientific visualization has brought us some tangible advantages.[25] But let others write extensively of these gains, for here we are concerned more with the future of visualization. Visualization is becoming quite realistic. We speak not just of virtual reality experiments but also of the visual modeling of dynamical systems. Graphical resolution and color surface rendering are better than ever, and most important, the connection between computation and graphics is becoming tighter. It is now possible to model whole molecules, have them behave according to the laws of quantum chemistry, and in this way start to "do chemistry" in some sense. Here is a specific example of the importance of multimedia to science: It is hard to imagine certain modern medical drug designs, for which molecules themselves have been analyzed through visualization, existing without the benefit of sophisticated multimedia tools for the chemist (and even, some day, for the doctor). In fact, versions of this kind of "multimedia-based science" are emerging in almost every scientific field. How many lives will be saved through the visualization of earthquakes, as a multimedia mix of geographical images, raw numerical data, graphs, sounds, and animation? And what will be the future state of the scientific field of seismology? Our basic point is that there is no telling what will become of the fundamental scientific fields on the basis of multimedia backlash.

It is often said that science—especially theoretical science—aims to provide a description of the real, physical world. Whereas virtual reality researchers may be concerned with *simulation* of the physical world, the goal of multimedia-based science is *description*. Imagine—and this is very hard to do—advanced multimedia several decades from now. But it is not so hard to imagine the multimedia technology coming ever closer to precise descriptions of reality. We might imagine theoretical chemistry being done in some kind of window system that also allows a tie-in of the theory with laboratory visualization of chemical experiments. Thus, future multi-

---

[25] Let us consider just three of the myriad valuable examples. In the 1960s, particles known as "solitons," which thrive in a world of nonlinear differential equations, were actually discovered by graphical modeling of these equations. More recently, erudite constructs such as the celebrated Mandelbrot fractal were discovered via graphical output. Even more recently, advances in medical drug design have taken place in experiments to which visualizable media are integral.

media-based science might be a *mélange de genres,* one manifestation of whose harmonious mix is the feedback between theory and experiment.

Before pursuing further the notion of multimedia-based feedback between theory and experiment, let us digress with an abstract metaphor. Imagine a roiling and angry ocean. Dark, nonlinear waves develop cusp structures in seemingly random fashion. These cusps are always in motion, ephemeral. One sees here and there a whitecap, the signature of excessive nonlinearity. Every so often a cusp is sufficiently singular—or conspiratorially collides so violently with another cusp—that a bit of foam is ejected into the atmosphere. This is a kind of dynamical catastrophe—a discontinuity brought on by extreme nonlinearity. Let the roiling sea stand for Western science as we know it, which has undergone perpetual change throughout recorded history. We might think of the era of Newton as a catastrophe—not necessarily destructive, but certainly singular. Another catastrophe might be the advent of quantum theory in the early twentieth century. A telling question is whether, at the threshold of the twenty-first century, the current obsession with computer visualization is just such a catastrophe. Or is it something worse: a signal of the very *demolition* of science? One reason why the great catastrophes of history might some day pale in comparison to a multimedia catastrophe is that visualization threatens to invade each and every scientific domain. It is this very universality of the visualization catastrophe that could make the difference. If we are right about universality, then the *mélange obscur* had better not prevail, or woe be to science.

Let us look at a telling observation by Clark.

> The image the rest of the world has of physicists, and even the image we have of ourselves, is usually a decade or two out of date. The entrenched images tend to be the astronomer glued to his telescope eyepiece, the particle physicist peering into his cloud chamber, the theoretician working with his equations on a blackboard. These images are memories of a bygone era. Increasingly, if one asks a physicist how he spends his research time, his answer, though he may hate to admit it, is that he spends it communing with a computer.[26]

These remarks amount to a caricature: Many physicists, and scientists generally, would object to the "bygone era" phrase. But what is most disturbing is the trend Clark describes. The tone suggests that perhaps the computer is actually doing physics—that it is at best a companion to the

---

[26] Clark 1989.

physicist, at worst a new breed of physicist. Can a computer do science, and if so, is visualization at all relevant to this question? We can answer the latter question immediately: Yes, visualization is entirely relevant, because it is through visualization that machines provide the most apparently realistic models of nature.

We do not believe what might be termed a wholly nostalgic view to be tenable. Nobody can deny the intrinsic charm of those antique eyepieces and cloud chambers, but we want to avoid such issues of aesthetics and soberly address the question of whether modern computing machinery can actually do science. We believe there is no *a priori* reason why machines cannot, in principle, perform feats of actual science, even though it may be argued that this has not yet occurred. Consider the following hypothetical example from high-energy particle physics. Imagine a physics supercomputer, running in the next century, that after analyzing reams of particle beam collider data and astrophysical observation data and performing the symbolic manipulation of quantum field equations, announces, "Tachyons do not exist."[27] Now *that* would be a genuine physics result, of the same stunning type we have learned to expect of human theorists. To imagine a wider range of examples of possible scientific performance by machines, let us turn to the notion of visualization and multimedia combined with actual scientific pursuit.

Here is how visualization is changing the march of science, an example of how a scientific effort can take the form of more and more advanced phases of visualization. One of the things that is still not very well understood by scientists is the liquid state of matter. The reasons for our insufficient understanding are legion.[28] Imagine a graphical model that displays the behavior of a liquid on the basis of some new theoretical model that scientists hope will be effective. A human observer of this visualization might look at, say, a moving waterfall made of this new, theo-

---

[27] A tachyon is an as-yet-mythical particle that always travels faster than light. It is perhaps one of many possible entities whose existence or nonexistence may well be settled eventually via combinations (symbolic, numerical, graphical) of computing modes.

[28] There are various intriguing ways to look at this failure of modern science. For one thing, a successful theory must explain why liquids such as water are virtually incompressible even though the molecules are loose and swarming. Another difficulty is the statistical decorrelation with respect to distance: Liquids have some properties of coherence (such as flowing in rivulets), but liquid molecules separated just a little exhibit basically uncorrelated motion. Even at the macroscopic level, the equations of gross fluid flow are quite unwieldy, even on supercomputers. Overall, one might say that the degrees of freedom for liquid ensembles are just too overwhelming. All these difficulties make for interesting—and very difficult—computer modeling problems.

retical liquid and simply decide that it "looks wrong." Now if a machine were really good at model computation, visualization, and pattern recognition, then one could imagine a model of the liquid state worked out entirely by machine. The machine would try ramifications of the model, "self-visualize," and then decide on the model's efficacy. This example is fanciful, but what it implies should be clear enough: The world of multimedia may evolve over the next century to the point where it deals a fatal blow to science as we know it. Perhaps more important, this might happen without any explicit perversion of the universally understood scientific method. Right now we see molecules colored (for convenient visualization) like Christmas ornaments, architectural structures with all the strains and stresses also visualized, multimedia models of working electrical circuits and chips, and so on. These are just practical visualizations, referred to in both the academic and the engineering literature as simulations, models, or virtual laboratory phenomena. But when machines begin to self-visualize, that might represent a radical cusp in the ocean of science.

Perhaps the future of multimedia will be a constructive one, something quite beyond an isolated graphical display. One can envision a machine future in which solid objects would be modeled and grown with ease, medical drugs would be designed proactively in advance of the epidemic, and we would thrill to full-length movies where all the human faces (and the psychological forces acting on them) were synthetic. The future of multimedia, then, will depend on how closely we expect machines to approach physical reality. And we are quite sure that a *mélange obscur* concurrent with that asymptotic approach bodes ill for the future of science. It may be none of the oft-mentioned negative cultural influences—such as numerical analysis, graphical games, and approximation models—but rather multimedia investment itself, that will ultimately bring about the demolition of science.

References

Augarten, S. 1996. *The Invention of Personal Computing,* unpublished manuscript.

Clark, B. 1989. *Computers in Physics,* July/August.

Davenport, G. 1996. "Indexes Are 'Out,' Models Are 'In.' " *IEEE Multimedia,* Fall, pp. 10–15.

Landauer, T.K. 1995. *The Trouble with Computers.* Cambridge, MA: M.I.T. Press.

Macaleese, R., and Green, C., eds. 1991. *HYPERTEXT: State of the Art.* Oxford: Intellect.

Subrahmanian, V.S., and Jajodia, S., eds. 1996. *Multimedia Database Systems: Issues and Research Directions.* Berlin: Springer-Verlag.

A NETWORK ORANGE

> The myth of technological and political and social inevitability is a powerful tranquilizer of the conscience. Its service is to remove responsibility from the shoulders of everyone who truly believes in it. But in fact, there *are* actors![1]

UNPREDICTABILITY

The more one thinks upon it, the more singular the modern network looms, especially if one imagines oneself in the armchair of a pre–World War II radio listener. Back then, communication was patently asymmetrical. Though the public rejoiced in the new radio information conduit, the radio network did not support interactivity. Connectivity per se was relegated to the telephone. And not everyone could telephone President Roosevelt. What we might call a modern "network orange" is a prime example of the intrinsic potential for chaos in technological growth. By chaos we mean neither pandemonium nor unholy purpose. The network is chaotic in the modern analytical sense of unpredictability: Who could have known what the network would become, and how can we possibly assess the effects of whatever forces are shaping its future? In this essay, we begin with some remarks on what queer clockwork the network has become

---

[1] Weizenbaum 1976, pp. 241–242.

and then turn to the issues of responsibility: what people now do, what they can do, and what they perhaps ought to do with the network.

Just like certain technologies discussed in Essay 1, the modern network is undeniably a military product. The grand Internet and perforce the World Wide Web were spawned by an earlier network manifestation, itself directly motivated by military interests.

In fact it was the Department of Defense (DOD) that championed ARPANET as an interconnection medium for government-sponsored research. This in turn led to the concept of an academic network, eventually to connect virtually all colleges and universities in the world, not to mention a great many corporations and governmental installations.[2]

Perhaps the armchair observer in 1940 could have predicted the Internet/Web, but who could have predicted the eventual rise of chat rooms, the posting of political opinion, or the marketing of lingerie? Even the idea of anything like electronic mail (with its bidirectionality) or teleconferencing (with its interactive video plus audio) would have been a stretch for the radio listeners to whom Roosevelt delivered his fireside chats.

But the network's assumption of its present scope and character is not its only chaotic aspect. Consider its topology. It is a mistake to think of the network as a vascular system, as similar in structure to arteries and veins. Another way to say this is that the network structure is not a typical "tree" structure. The image of trunks, from which branches emerge, from which twigs and finally leaves (users) emerge, does not capture the network's organization and growth. These more closely reflect a true network, or "neuronal," topology.

An interesting and, for our purposes, a telling biological analogy comes into play here. Vascular systems involve the transfer of energy within an organism. In a mammalian system, both oxygen and glucose are conveyed via arteries. But neuronal systems involve the transfer of information. One has to learn time and time again—whether in biology, physics, or computer science—that energy and information are not the same. Ergs and bits might as well be apples and oranges.

Sure enough, the topology of arteries and veins is not the topology of nerves. A vascular tree structure involves the transport of energy to every place that needs it. Virtually every cell is reached via capillaries, at the bottom level of the vascular tree (technically, below the final capillary level there is a local perfusion: a "bath" of blood derivative). Evidently, one tree

---

[2] See Rheingold 1993 for the interesting history of these developments.

mechanism suffices for delivery of both food and oxygen. The vascular system stands in for the ancient environment of our single-celled ancestors. Single cells living three billion years ago were directly exposed, by virtue of *their* topological structure, to the environmental bath: A single cell's membrane interface was sufficient. Much later, the association of cells into organs necessitated the transport of energy. A mammalian liver cell, for example, obtains oxygen ultimately through a complex lung-blood interface. The cell is essentially at the bottom level of the vascular tree.

By contrast, although some information is transmitted chemically through the vascular tree—mainly via the endocrine system—this information transfer is diffusion-based. Molecules are released and later employed. Again the vascular tree is the conduit. But evolved neural cells, such as those of the central nervous system (CNS), require much more intensive transfer of information. In fact, when neural cell density is high enough, those cells evidently require a *network*, where interconnectivity is the rule. To convey an idea of scale, the human system of sensory input/motor output involves 10^7 to 10^8 neurons and is in large part tree-based. But the CNS—including the brain with its staggering 10^{10} to 10^{11} neurons—is a network structure *par excellence*. Each neuron is in contact with about a thousand other neurons, not all of them close by. Such network topology is probably important for both accuracy and redundancy in an information-based system of high integrity.[3]

Thus the "network orange" has not grown in the simple way a tree grows by sub-branching upon branches. End users (cells) are not necessarily supplied via one defining branch (capillary). It is more accurate to think of some tree-like growth, supplemented in various ways by patching, overlay, supplementary growth, and cross-linkage.

Let's look at an example of overlay and cross-connection. At the Center for Advanced Computation where this essay is being written, a certain World Wide Web request might travel over the Internet, first going by optical fiber to a "Point of Presence" in downtown Portland, Oregon. The

---

[3] These theoretical considerations amount to speculation in the absence of any really thorough understanding of neuronal systems in general. (Witness the state of disrepair we attributed to AI in Essay 2.) And one can even go further. For example, consider this interesting question: What is the "fractal dimension" of the Internet, and does this dimension imply a level of strangeness or a true manifestation of chaos? (The fractal dimension would be obtained, say, by assessing the connectivity of users within a given geographical—or, better, a given demographic—radius, as a function of that radius.) Still, we feel that the issue of user responsibility expressed at the end of this section overrides any scientific responsibility, so speculation should be welcome.

signals are then routed to a generally busier node in Seattle, Washington. From there, they may be sent out of the Pacific Northwest to a corresponding node in Michigan. By now, nodes are in many cases owned and operated by corporations, but a variety of for-profit and not-for-profit concerns are involved. At the same time, back in the Portland area, there are other networks running. Without much trouble, one can access various, "overlaid" networks. What is especially conducive to future chaos is the potential for interconnection of telephone, cable television, and who knows what further networks of the current overlaid topology.

What renders the analogy to the nervous system even more telling is the impending cross-linkage of television cable and satellite systems, which are in principle available, along with the overlaid network, to a typical "cell."

Complex topology is not the only fuel for chaotic growth. It also matters what form of information the network can sustain. It remains difficult to effect teleconferencing and interactive video, and it is still difficult to handle sound on the network. But bandwidth limitations have always been overcome, and these multimedia bandwidth barriers will certainly fall, probably within a decade or so. Even now, high-performance network manifestations such as the Advanced Technology Demonstration Network (ATDNet) are being tested.[4] New bandwidth freedom will alter the basic topology of the network, and this should further augment the chaotic growth.

One recently emergent network phenomenon deserves special attention. Networks are beginning to show signs of self-organization and self-modification. It is becoming progressively easier for a new network node, whether for a college campus library or a new corporation, to declare its existence on the network. Parts of the network orange are spawned easily, and they immediately cry out for electronic attention. The network, in turn, is evolving response reflexes. For ATDNet, for example, analysts have described "ATM autodiscovery," whereby any new nodes are discovered automatically so that the full network undergoes continual self-organization.

---

[4] ATDNet was established by the Defense Advanced Research Projects Agency (DARPA) in 1994. This high-performance network experiment combines asynchronous transfer mode (ATM) with synchronous optical network technologies (SONET). The sophisticated mix of ATM/SONET is a step in the direction of high-speed, high-volume networking. See Bajaj et al. 1996. Of course, if ATDNet works out, we shall have to admit one day that yet another evolutionary tier was generated by military interests.

The difference between vascular and neuronal topologies implies a difference in responsibilities. We think it is important for the individual computer user *not* to imagine being the cell that receives sustenance of some kind and then performs some function. Multiple supply and feedback lines to a given cell exist now and will undoubtedly increase dramatically in the future. *This amplifies the individual responsibility of network participants and coherent groups of participants.* There is no telling what this prescription for chaos will bring. To say it in another way, the modern network user should try to imagine what future network scenario will compare to the current Internet as the Internet compares to 1940s radio. We thus turn to the problem of maintaining such notions as truth, responsibility, and integrity—if there can be maintenance in chaos.

ORACLES AND ACTORS

A galaxy of books, articles, and magazines have taken up the challenge of forecasting our computer future, a future in which almost every element of society, from individual lives to political and social policy, will be reshaped by computers. Depending on which publication we pick up first, we are either at the beginning or in the midst of a "computer revolution" or an "information revolution." If we just remain at our screen and keep up with the latest innovations, we can move with our laptop of choice into the vanguard of the greatest cultural transformation humanity has ever witnessed. Of course, counting the dollars we have used to get there will be a mite dispiriting, but the only rent we need pay to live in the new age is the cost of a PC and some networking. Apparently our network orange not only is being fed and allowed to grow (in some instances even to enlarge itself) but also is being aggressively promoted. Of course, every PC and its accompanying software are outmoded as soon as purchased, but keeping up with the upgrades is said to be well worth the price.

Because revolutions, especially heavily promoted ones, are by nature enticing and full of promise, it is tempting to believe the advertisements and eagerly buy in, both fiscally and philosophically. We ask the reader to resist this temptation and at least momentarily leave utopian interpretations of the "computer revolution" to news weeklies and the latest crop of futurists.

There is a tendency for those living in any given society to conceive of it as undergoing revolutionary dislocation and to see themselves as experiencing dramatic transformation. Society is teetering at one or another brink, the moral fabric has deteriorated or is being rewoven into a radically new garment, the changes are ubiquitous and overwhelming, the remedies drastic.

It is also commonplace to discover that the champions of social change are prone to self-serving characterizations of the magnitude of such change and of its long-range import. Those who are most concerned with social change are often in the worst possible position to judge its significance. We may be right to think of the late twentieth century as reflecting changes so rapid, so radical, and so sweeping that they amount to a "computer revolution." No doubt many feel a revolution is upon us, but we do not *know* it is a revolution, and it will be some time before we *can* know.

The kind of rhetoric we use to talk about ostensibly large cultural movements can also affect the way we behave toward those movements. When we refer to something as revolutionary, we make it by definition a pretty earthshaking force that can make some of the social actors involved overly exhilarated and some overly frightened. The first group tells us we must be tough-minded, face the facts, give up our romantic nostalgia for a past now irrevocable, and not only ride the new wave into the future but also turn our energies to directing it so that, wherever it goes, we will be there first. For their part, those who dislike what is happening think of it as inexorable and retreat into isolation. The danger is that we can become so busy trying either to lead the revolution or to retreat from it that we stop thinking.

But it would be irresponsible to proclaim the network either glorious or evil. We do not yet know what it will be, but we also should not believe that the growth of the network is a juggernaut and that our only alternative to climbing aboard is to get out of the way.

The BBS as Canonical Educational Testbed

Much of what we observe about user tendencies and responsibilities can be exemplified in one relatively old and simple network communication paradigm: the Bulletin Board Service (BBS). A BBS offers the essential ingredients of computer-mediated communication without the complication of commercial issues. Operating in campus environments with largely college-student (or at least college-age) users, a BBS would be expected to complement and even enhance the users' educational experience. Above all, the simple BBS model reveals to some extent *what people really tend to do and say on an available network.* All this makes the BBS a good canonical system for analysis.

A BBS is a software adjunct of a computer or mainframe system that enables other computers to call it up via network. Since its development in 1978, when Christensen and Suess wrote the first BBS software, there had grown up, by 1993, an estimated 60,000 BBS systems in the United States.[5] The idea, of course, is to provide news of interest to the BBS users. A popular BBS version allows interactive communication among users, or "on-line conferencing" (OLC), a special and very powerful form of communication by which messages can be exchanged without delay from user to user. As with a telephone, the user can be heard immediately and can get an immediate reply.

Many forms of communication analogous to the BBS are flourishing to some degree or other, and all are mutating into what will be (at least for a day) the very latest version: Internet Relay Chat (IRC), MUDs, MOOs, Talkers.[6] Their common element is their use of OLC, which computer jargon calls "synchronous chat programs." If we have had the experience of talking to a roomful of people, we can imagine what it be would like to type to a roomful of people and to have these people read and respond to our words as quickly as we get them out.

How well do BBS services, including OLC, accomplish their goals of providing effective communication and serving as a collective instrument of education? The BBS is often touted as a superb medium by which ideas and information can be purveyed, challenged, and adjudicated in an informed public setting. That surely sounds a lot like what we think education should be.

The Iowa Student Computer Association (ISCA) is probably the largest BBS in the United States, accommodating at any given time about a thousand users. ISCA describes itself as "a place where users can be

---

[5] Rheingold 1993, pp. 132–136.

[6] IRC is a multi-user worldwide network that enables users to converse in real time. It is distinguished by the use of channels, and the names of the channels reflect the interests to which they are dedicated. MUD is an acronym for Multi-User Dungeon. Unlike the BBS and IRC, MUDs customarily include adventure games and accommodate the building of new locations or "rooms." Because users can take on fanciful identities in their game playing, it is sometimes considered one of the earlier forms of "virtual reality." A MOO (Mud-Object-Oriented) is similar to a MUD, but allows more in the way of building objects and giving invented characters particularity. Talkers are comparable to the BBS but are pruned of everything except programs that permit OLC.

educated[7] and entertained by reading and posting messages about various topics or socialize over the one-on-one express message system and personal intra-system mail."[8] There is not much doubt about the entertainment. It is the purported educative function that we wish to examine. Because a computer is less expensive than college tuition, it is of some importance to consider whether keyboards, with the aid of OLC, can complement, or even improve on, the classroom. The question is whether the networking of users—even in the simple BBS mode—brings about or contributes to education, as education has always been understood.

The first thing to be said about education is that it necessarily involves the transmission of what is intellectually worthwhile. This intellectual content makes education markedly different from training, which can also transmit (if in a different way) what is worthwhile. Sales clerks are trained, not educated. They are trained in a specialty. "When educationists proclaim that 'education is of the whole man,' they are enunciating a conceptual truth, for 'education' rules out narrow specialism just as it rules out a purely instrumental approach to activities."[9] Education is a process aimed at communicating established connections to the learner. Therefore, it is directed to providing knowledge and to encouraging the integration of data from which knowledge comes. R. Pring puts it this way: "In any knowledge whatsoever there must be some sort of integration, of seeing otherwise unrelated events as events of a certain kind, of structuring our experiences by means of concepts."[10] In education, the emphasis is on structuring, connections, and integration in the mind of the learner. This is as much as we will say here about the distinction between education and training. We ask the reader to grant it provisional acceptance. The larger defense will appear in Essay 5.

We hope our dilemma is becoming apparent. The media are powerful, and along with our president and vice president, they glorify the network revolution and its positive impact on education. We cannot accept the idea that both education and training are natural by-products of the network, and we have to wonder how to get sober assessments of each as they occur in network communication.

---

[7] Our italics. There are countless examples of the claim that education is an implicitly "natural" by-product of networking itself.

[8] From the World Wide Web Home Page, 1996, of the Iowa Student Computer Association.

[9] Peters 1973.

[10] Pring 1971.

One career path that may cause a trained person to be less than ideally educated is the path of specialization. But specialization is the hallmark of network communication such as BBS chat. Which ideas arise and dominate on a BBS seem to be determined by whims of exquisite subtlety. Not only is the growth of the network orange chaotic, but so is the manner in which specialization appears in network conversation. It may well be possible to obtain training on some topic via BBS, but there seems to be none of the intellectual comprehensiveness or ordering of the world that could provide an educational outcome.

It is generally agreed that a network's function is the transmission of information. We think it is dangerous to overestimate the value of this task. Information—accumulated and organized data—is surely something to be valued, but *acquiring it is only one aspect of becoming educated.* Otherwise, we might as well hail a telephone book or a world almanac as the pinnacle of intellectual achievement. Moreover, because the network orange is based on digital technology, with its attendant "hard" definition of information as an assemblage of bits, even alternative notions of information become progressively more obscure.

We submit that regardless of the theoretical heights attained in the work of Claude Shannon,[11] to label virtually everything information is intellectual folly. Valuable alternative definitions of *information* include the properly expressed *relationship* between facts, the *quality* of those facts, and even their *relevance* with respect to other bodies of facts. We wish that network users would harbor—either personally or as a group—some notion of the quality spectrum of information.

But the network world flattens any such spectrum. Arrant nonsense, patent falsehoods, bad jokes, and Newton's laws of motion are equally promulgated as "information," as long as they are translatable into a binary code. Talk about information is talk about signals, and about the transmission of signals, whose semantic content has no bearing on their status as signals or on the integrity of their transmission. Therefore, $4 + 5 = 81$ is information as much as $7 + 5 = 12$ is information.

Then there is what we might call "information posing." We recently were appalled to see a colorful brochure describing the historical origins of a tea beverage. The brochure amplified its claims in this way: "It's true. You can read about it on the Internet, at [. . .site]," as if to imply that

---

[11] Shannon 1948. Shannon seems later to have regretted using *information* as a basic nomer in his theory; but it took hold, continues to be used, and continues to foster "blurring" of the quality spectrum.

electronic deposition of information is somehow self-validating. (Compare another phrase often included in print advertising: "As seen on TV!")

As the BBS is now deployed, little that is done to it or with it answers to any conceivable goal of education. Indeed, if one understands that education is neither simple fact-gathering nor undisciplined self-expression, the case is closed when it comes to evaluating the BBS as educational tool. The only open question is not whether the resources of a BBS contribute in some positive way to the process of education; it is how much harm they do.

LANGUAGE MANGLING

Almost every BBS possesses four features: an accessible database, ranging from highly technical scientific information to airline schedules and stock market quotes; a collection of file servers with such things as shareware software accessible to the subscriber; a set of public forums in which contributions, "postings," are invited on almost every conceivable topic from "Macintosh" through "Peace" and "Human Nature" to "Sex" and "Puns;"[12] and the glamour item of the BBS, an interactive OLC mechanism that permits nearly immediate communication among users.

Of course, there seems nothing particularly attention-getting about the equivalent of a telephone receiver for written text. But BBS is a radically different instrument of communication. The conventions of our society frown on using the telephone to dial up complete strangers and engage them in conversation. Only obscene callers and telephone solicitors do that. The conventions of the inner BBS society, however, validate such acts and even condemn as "lurkes" those who are loath to participate in them.

The random-access, random-connect phenomenon is not just an aside. The general canons of conduct, thinking, and language in a BBS are derived from the agreed-upon conditions under which message giving and responding are allowed to function. It would even seem that the interactive mechanism in a BBS provides (1) a foundation on which faceless denizens of an insular technological world can build human relationships and (2) the conditions of socialization and even camaraderie that make that world habitable.

---

[12] Here are some titles of forums from a typical BBS: "Babble Kickass Poolhouse," "Neverending Limerick," "Netsex and Pull My Finger," "XRated," "DOS and Windows," "Sports," "Creatures of the Night," "Star Trek Galaxy," "Cyber Alternative Music," "Heavy Metal," "The Muppets," "Philosophy," "Star Wars."

The remarkable and oft-abused fact about BBS is that its interactive mechanism provides instantly—without duplicating machines, xerography, or endless verbal rehearsal—the public environment where users can be initiated into the process of mastering themselves and the world.

Yet this mechanism has contributed to no such result and has even created a network of evaluative conventions that are completely at odds with the aims of education. These conventions are reflected in the peculiar, astonishing language of the BBS. A. Huxley once remarked that "conduct and character are largely determined by the nature of the words we currently use to discuss ourselves and the world around us."[13] If he is right, as almost surely he is, the great potential of the interactive mechanism of the BBS has not been fulfilled. It fails to function as a public arena in which the cognitive content of ideas is subjected to scrutiny and evaluation. Something else has taken over, and if it is communication at all, it bears no resemblance to the forms of communication that contribute to education. Instead, cartoon words are used to carve out cartoon facts. There is nothing wrong with that. Most of us enjoy reading the comic strips, but we don't delude ourselves that they contribute much to our education.

The ostensible discussion of ideas on a BBS is not, in any genuine sense, a discussion of ideas at all. Nor is it the automatic product of its underlying technology. As the users of a BBS employ it, interactive communication becomes, and requires, immediate communication between caller and responder. Remember that the communicating users have no names. They are identified by aliases that the BBS calls *handles,* a term presumably borrowed from truckers who communicated with each other on CB radio. (Much of the truckers' message traffic had to do with avoiding speed traps, so there was an advantage in anonymity.) There is also a major difference between truckers and the users of BBS. When truckers used CB, they continued working. Therefore, the trucking industry has nothing comparable to the Internet Addiction Support Group (IASG). The group provides on-line counseling for computer users who are jeopardizing their livelihoods, their academic careers, and even their lives through obsessive use of Internet resources.[14]

The BBS world, like any other world, is governed by conventions, and as in the larger world, the conventions that are unspoken are the most

---

[13] Huxley 1962.

[14] Using the resources of IASG requires access to the Internet. It is exactly as if Alcoholics Anonymous scheduled its meetings at the local bar.

revered. One of these unspoken conventions is that the status of a user is a function of the number of users with whom she or he has established relationships, and of the intensity of the relationships. Success in establishing as many relationships as possible is an inverse function of the length of messages sent and replied to. Every message from a skilled—therefore esteemed—BBS user will be short, crisp, conceptually skeletal, and factually empty; this last property is a function of (and is protected by) the disguise mechanism, the use of handles. If a message does not elicit an immediate response, the receiver will fear that the sender will turn to another. And the status of the dilatory receiver will have been diminished.[15]

Messages must at all costs avoid complexity or qualification, which require more extended exposition and therefore risk the possibility of a tardy reply. No penumbra of obscurity or vagueness can hover over the content. Either will require expansion and, therefore, too many words. The paradigmatic message will be a quip, a slogan, or a snappy, pungent phrase; the ideal response is quip to quip, slogan to slogan, phrase to phrase. It is the language one could expect in a singles bar, which is no surprise given that, as Christopher Lloyd has remarked,

> Computer networks are the singles bars of the 1990s. The mysterious world of computer messages, known to some as electronic mail, to others as bulletin boarding, and generally as cyberspace, is now one of the most dynamic, but least known, ways that people meet often in secret and exchange intimate conversation. Every night, thousands of lonely hearts in Britain, America and all around the world join people who simply want to "chat."[16]

Network chat is a form of communication in which the complexity and richness of the world and the self are reduced to formulaic utterances. In this sense, the language of chat is precisely antipathetic to the aims of education. If education is the initiation of students into different forms of knowledge (science, history, mathematics, religion, aesthetics, and such practical types of knowledge as go to moral and prudential forms of

---

[15] In an extended and incisive study of IRC marred only by some unnecessary lapses into the latest sociological jargon, E. Reid comments, "Speed of response and wit are the stuff of popularity and community on IRC. The Internet Relay chat and such social endeavour demands speed of thought—witty replies and keyboard savoir faire blend into a stream-of-consciousness interaction that valorizes shortness of response time, ingenuity and ingenuousness in the presentation of statements. The person who cannot fulfill these requirements—who is a slow typist, who demands time to reflect before responding—will be disadvantaged." Reid 1991.

[16] Lloyd 1993.

A Network Orange

thought and action), and if this initiation requires differentiated and complex awareness of the canons by which judgments can be understood and ratified, then the user has agreed to occupy the most ill-informed of worlds. It is not only a radically simplified world but also one whose conversational strategies are determined by the fact of social "cluelessness," a lack of the cues that in face-to-face conversation are provided by such things as facial expression, body language, and perceived social status.[17]

The conventions also mandate a form of discourse in which civility can be supplanted by acrimony, and wherein mutual insult can graduate to virtue. This discourse is called "flaming" or, more technically, "disinhibited behavior." An effective flamer is evidently a person to be reckoned with on a BBS. No one who disagrees with an expert flamer is safe, and the response of the flamer will be swift and deadly. Indeed, users look forward to an especially lively episode of destructive prose. For some, there is nothing quite so stimulating as a flamer launching into the kind of exchange associated with snapping towels in a high school gym class.[18] E. Reid, who on the whole appreciates what she calls the "postmodern" aspects of computer communication, admits that "on its more negative side the disinhibiting effect of computer-mediated communication encourages the expression of dissent, rebellion, hostility, and anti-social chaos."[19]

Even in the BBS forums or rooms, where views are posted and replies are appended later (the interactive feature thus not operating), there is the same general insistence on quick items quickly spoken. Almost never is any problem or subject, however complex, addressed in more than ten sentences, and responses usually comprise much fewer. The following are some actual postings from a BBS forum entitled "Philosophy."

> [Poster A] fine yes, and i admitted my flame was human. That's that. the room info has easy rules. . real easy. think before you post. its easy. i expect

---

[17] The classic studies of cluelessness come from D.R. Rutter. See, for example, Rutter 1984.

[18] Lea et al. argue that the number of episodes of flaming has been exaggerated in the social scientific literature and that too much has been made of their significance. In so doing, they neglect the degree to which the perceived possibility of flaming, however uncommon its actual incidence, can affect the behavior of all. It is like arguing that the social importance of violence toward women has been exaggerated because there are fewer cases of it than was heretofore believed, a position neglecting both the impact of the violence when it occurs and the efforts made by frightened others to avoid it. Once a bully has established himself on the playground, he does not have to beat somebody up every day to maintain control and sustain fear. See " 'Flaming' in Computer-Mediated Communication" in Lea 1992.

[19] Reid 1991.

you all to do that. .if you don't, i do intend to flame. .or maybe you want i should just kick like other ra's [Room Aides] do? no, i think flaming is better in my opinion. . but vote seems to be for kicks. . .so that's the new policy if you wish. existentialism is the evolution of christianity. through the elimination of any god, it shows how we are thus god (ref.mad man in market square). obsessed with our mortality? hm hmm sounds sartrian. one tiny speck of the entire philosophy. well not tiny. .but still small.

[Poster B] Existentialism is the evolution of Christianity??? I don't give a rats ass whether that's your "specialty of your major" or not. Back it up. I happen to know a bit about existentialism, and from what I perceived, [F] is in the right here. So what if you hold yourself to be the highest being in your existence? Do you feel so insecure in that position that you must then call yourself a god? At this point one must ask what the word "god" means. In my vocabulary, god is the ultimate being, who knows all, sees all, and creates all, and, incidentally, does not exist. I consider anyone that believes in god is not realizing the full potential of their existence. So why ruin a good thing by calling yourself a god? To me it seems to be just a worthless title to give yourself to make you feel important, as lawyers do with "esquire". . . . Get a grip.

[Poster D] Doesn't it seem ironic that life never lives forever and death never dies?

[Poster E] Aha! I've found the cause of this disagreement. As you stated, the metaphysical structure of the universe proposed by almost everyone is that god is at the top, single and unified, while multiplicity is at the bottom, diverse and unconnected. I hold a slightly different view. I think that multiplicity is both at the top and bottom. . . It's complicated, what I believe, but It's kinda like that famous enigma: If a bunch of people in a stadium gathered around a tree, and all of them believed the tree was on fire, would the tree actually ignite? You see, I believe it is the thoughts of all of us combined that is the 'creator' of us. So if everyone in existence did believe without a doubt that the tree was on fire, then it actually would ignite. I can explain more about this theory if you're interested.

[Poster B] *yawn* enough with the God question, OK? I won't even try to answer it, since I don't believe in any God, or an omnipotent being, for that matter. . . However since I believe that our combined thoughts define what is, thus making us as a whole omnipotent, we have already created the rock we cannot lift. Except in our case it isn't necessarily a rock that we can't lift, since we are capable of designing equipment which can lift most anything, its more like a place we cannot go or a speed we cannot reach, etc. Since these limits exist, we believe in them, and vice versa, so we cannot overcome them. Personally, though, I think we will eventually find a way to go faster than light. . . Anyone have any philosophical thoughts about art/music and such? Sounds like an interesting topic to me. . .

[Poster F] I must say that I disagree about Pee-Wee's use of underwear being an expression of his perception (or lack of perception) of underwear as headgear. Clearly he was addressing the sense of the comic that we all seem to attribute to underwear due to our sexual insecurities. The child that Mr. Herman represents for us recognizes the need for levity in sex-issues. Therefore, his use of underwear is not so much a reenforcement of his own perceptions as an attempt to make us more comfortable with ours.

[Poster G] You are wrong. . . . I do recall PeeWee Herman using underwear as a form of headwear. So in this case, PeeWee perceived underwear as a hat as his truth. I was also talking about dirty underwear and socks. . .what's one person's truth about dirty is another person's truth about clean. Ummm does this make sense? Any one who doesn't consider this philosophy can eat what's under my bed.

All the foregoing postings are from a forum where serious discussion might be expected. The subject is philosophy, not heavy metal. However entertaining the contributions may or may not be, they are, from the point of view of serious thought, puerile.

If the form and content of language use affect our understanding of the world and what it consists of, then chat gives us a world shaped by the necessity that BBS communication fit into the tiniest of conceptual pigeonholes. Whatever does not fit will either be squeezed into shape or declared irrelevant. Serious inquiry expressed in serious prose would take too much time and would probably not be read, because that would disturb the BBS rhythm of impromptu assertion and quick, unstudied response.

When it comes to emotion, where the expression of rich nuance has challenged our greatest novelists and poets, chat has pared the infinite spectrum of feeling down to an impoverished set of conventionalized expressions, usually no more than five or six. Further, consider that the language that seeks to catch the special character of a feeling exists in a domain where facial expression, action, gesture, and attitude are intimately bound up in the task of getting the feeling right. Not so in chat. Its conventions eschew the complications and ultimately deform the feeling.

The word conventions in chat are amusing—and revealing. Emotion-words must be set off by asterisks, though, as in our hurry to read a message, we might otherwise not notice them: *grin,* *smile, smirk,* *frowning,* *looking reflective,* and *knitting eyebrows.* In fact, fewer such words are usually deployed. And if proper expression would take too much time, a set of semi-graphical icons is available. The classic smiley face becomes : −), sadness turns into : −(, and distress or disappointment is conveyed by : −/. One reporter puts it this way:

> Many of their conversations were in a clandestine code known as "smileys," a language that uses symbols as a kind of shorthand to express emotions. The widespread use of the smiley code was highlighted in last week's issue of *New Scientist* magazine, which, in a short feature on computers as Cupid, provided its readers with a selection of the most useful symbols. Each is read by tilting the head 90 degrees to the left. For example, : −) is the original smiley that cyberspace chatters put at the end of a joke, a trend that began about three years ago. Recently, the language has grown to a vocabulary of thousands of symbols. Here are a few of the more bizarre examples: ! −(means "blackeye"; $−) means "yuppie"; %@:−(means "hung over"; &.(. means "crying"; (:+) means "big nose"; : −E means "buck-toothed vampire."[20]

It is nearly intolerable to think of a world in which these words and simplistic icons "canalize" (in effect, mold) feelings and, more tragically, begin to reflect them. We must remind the reader that we do not think this result is inevitable. The BBS might be used differently, but that would require recognition of how badly it is used now.

THE EMERGENCE OF THE WORLD WIDE WEB

The World Wide Web has already become, for many users, a network manifestation of choice. The workings of the Web are not hard to understand.[21] It is a collection of pages and links to pages, some personal, some commercial, and some organized by theme such as poetry, cosmology, or genealogy. Though the Web is quite a recent manifestation, the same concerns and cautions that we voiced in regard to the old, canonical BBS apply. The very structure of HTML documents is biased toward the transfer of textual information. Web pages offer greater freedom in presentation, but limitations abound. As with all prevailing forms of network traffic, there is still no good way to present information in what might be

---

[20] Lloyd 1993.

[21] Briefly, the Web involves the maintenance of and access to hypertext (HTML: hypertext mark-up language) "pages" comprising a multimedia mixture, say text and graphics, with efficient means of manually following "links" to other topics or even other Web "sites."

called a holographic or even holistic fashion.[22] One is forced to follow "links," and this constraint favors specific kinds of information layout. In this respect, a Web site is similar to a sophisticated table of contents.

A site is, however, a very special table of contents, in which both categorization and substance in the theme-pages are beyond evaluation by the nonspecialist. We have spoken of specialization before, and the Web is a particularly good example of how the need for expert knowledge collides with the form in which the Web presents its theme-pages. If we wish to know about a field in which we are not experts, we have to depend on a prior evaluation of its presentation by an expert. We lack the knowledge to do it ourselves, and the Web does not do it for us.

T. Oppenheimer provides a cogent warning about the reliability of "the Net."

> On almost any subject the Net offers a plethora of seemingly sound "research." But under close inspection much of it proves to be ill-informed or just superficial. . . . Even computer enthusiasts give the Net tepid reviews. "Most of the content on the Net is total garbage," Esther Dyson acknowledges. "But if you find one good thing you can use it a million times." [Stephen] Kerr believes that Dyson is being unrealistic. "If you find a useful site one day, it may not be there the next day, or the information is different. Teachers are being asked to jump in and figure out if what they find on the Net is worthwhile. They don't have the time or skill to do that." Especially when students rely on the Internet's much-vaunted search engines. Although these tools deliver hundreds or thousands of sources within seconds, students may not realize that search engines, and the Net itself, miss important information all the time.[23]

The Web makes it easy for ignorance to parade as knowledge. Pages in the Web, however arcane or sophisticated, come to us without the provenance of expert endorsement. Little on the Web comes to an innocent user with unassailable credentials, and a user is at a loss to determine what might be relevant, let alone true. Links do not help, precisely because it is the author of the home pages who does the linking. Although divisions of knowledge have become more and more specialized, the Web behaves as though all its users were polymaths. Furthermore, it fosters the egalitarian idea that having acquired expertise in hypertext generation and some skill

---

[22] It should be said that some auditors of the network have found therein the perfect, if revolutionary, holistic presentation, redefining *text, authorship,* and *reading.* Some of the intellectual acrobatics involved in doing this are revealed in Tuman 1992.

[23] Oppenheimer 1997.

in typing is somehow akin to having published a book or a paper. There is something to be said for home pages as an international vanity press, but they leave the nonspecialist reader without much guidance in judging the cognitive worth of what he or she reads.[24]

The distinguished intellectual historian G. Himmelfarb makes the point brilliantly.

> But democratization of the access to knowledge should not be confused with the democratization of knowledge itself. And this is where the Internet, or any system of electronic networking, may be misleading and even pernicious. In cyberspace, every source seems as authoritative as every other. . . . The search for a name or phrase on the Internet will produce a comic strip or advertising slogan as readily as a quotation from the Bible or Shakespeare. The Internet is an equal opportunity resource; it recognizes no rank nor status nor privilege. In that democratic universe, all sources, all ideas, all theories seem equally valid and pertinent. It takes a discriminating mind that is already stocked with knowledge and trained in critical discernment to distinguish between Peanuts and Shakespeare—between the trivial and the important, the ephemeral and the enduring, the true and the false. It is just this sense of discrimination that the humanities have traditionally cultivated and that they must now cultivate even more strenuously if the electronic revolution is to do more good than bad.[25]

Some pages on the Web are beginning to acknowledge the problem. One group that calls itself "The Mining Company" advertises its service this way: "Take back the Net. Hundreds of real guides mining the net for you."[26] The purpose is to come to the rescue of the untutored user who encounters a host of *ad hoc* links on some subject and cannot devote a day to following them out or a month to evaluating them. Competent guides, presumably specialists, lead the user through the morass. Conscientiously applied, that might be a promising solution to the anarchy of endless unevaluated links on the Web.

---

[24] Here is some salutary wisdom from Dr. John Grohol: "If you're looking for additional information on any particular [mental] disorder online, you're probably in the wrong place first. Online is great for ease of use and fast, instantaneous results. It's a wonderful place for support groups and meeting lots of people with similar problems throughout the world. But it's a lousy place for detailed and well-written information. . . . In-depth information on any disorder's causes and treatment is not to be found online. Your best bet, first and always, is still either a large retail bookstore or a large university library. . . . I know it's more work on your part and all, but it's the truth of the current state of affairs on the Net today" (http://www.coil.com/≈grohol/faq.htm#question_1).

[25] Himmelfarb 1997.

[26] http://www.miningco.com/.

Nonetheless, the very nature of the Web and the conditions of its reading are more conducive to gathering facts than to organizing them into an interpretative or theoretical schema. In calling attention to a crucial difference between reading a book and reading a screen, Himmelfarb offers some powerful reasons why the Web is better suited to providing facts than to promoting thought.

> Holding the book in hand, open at that, we can readily concentrate the mind upon it, to linger over it, mull it over, take as long as necessary to try to understand and appreciate it. Reading it on the screen, however, is a quite different experience. There we tend to become postmodernists in spite of ourselves. It takes a great effort of will to concentrate on the text without regard to whatever else may happen to be called up on the screen along with it. And it takes a still greater effort to remain fixed on a single page without scrolling on to the next, let alone to concentrate on a single passage, line, or word. The medium itself is too fluid, too mobile and volatile, to encourage any sustained effort of thought. It makes us impatient, eager to get on to the next visual presentation. And the more accustomed we become to the medium, the more difficult it is to retain the old habits of thought. We become habituated to a fast pace, an ever-changing scene, a rapid succession of sensations and impressions. We become incapacitated for the longer, slower, less feverish tempo of the book. We also become incapacitated for thinking seriously about ideas rather than massing facts. For the purpose of retrieving facts, the Internet is enormously helpful, although even here some caveats are in order. We need to concentrate our mind on exactly what it is we want to know, to resist being distracted by fascinating but irrelevant facts, and—most important—to retain the ability to distinguish between facts and opinion, between reputable sources and dubious ones.[27]

Yet the Web is a strong step in at least one right direction. Web users can be more discerning with respect to what we have called the information quality spectrum. It is sometimes easier to ignore, via the hypertext linkage, useless or unsavory pathways, though without relevant knowledge it is also easy to be lost in such pathways forever. It is likewise easier to jump over whole sections or sites, which is very hard to do with BBS or rudimentary e-mail systems. Then, too, there is the ease with which a Web site can be modified and improved. For these reasons, the Web may evolve under a more stringent selection pressure than previous network systems. Given some sense of responsibility in users, this selection pressure is beneficial: Certain properties of the "network orange" should

---

[27] Himmelfarb 1997.

indeed become extinct when more useful properties arise to compete with them. And nothing hastens extinction like selection pressure.

ON THE ISSUE OF NETWORK RESPONSIBILITY

Finally, we address the issue of responsibility. The BBS, the World Wide Web, and their successors could be converted into vital and purposeful educational instruments. Readers who wonder why education should be a chief preoccupation—beyond the obvious concern of the present authors—should recall the academic origins of the "network orange." It would be tragic to lose that original perspective entirely. We submit that educators must accept the responsibility of giving the network clear objectives and of ensuring that the network is able to support those objectives.

Educators who neither worship nor abhor technology can help convert a technological artifact into a functioning tool, and their cooperation during network evolution could be of the very greatest educational importance. It should be possible to convert the older, interactive BBS paradigm from something like a technological dating service into more of an electronic village with appropriate information categories that, among other functions, somehow suppress the *ad hoc*. Interesting attempts to effect such a transformation have emerged,[28] but it is still not clear how the Web should evolve into a fully interactive medium along such lines. Some authors have pointed out that the current Web "fails to mirror the functionality of today's paper-oriented learning environment" and have suggested more advanced forms of multimedia browsers.[29] We applaud this kind of sober comparison between the old and the new.

It is not only educators, of course, who should practice network responsibility. The neural structure and chaotic properties of the "network orange" should serve to amplify every network user's sense of individual responsibility. One can only hope that network users will not rush headlong to use new network tools unthinkingly. We speak neither of censorship nor of control. Our wish is simply that the tools be thought of as finely edged ones. And our hope is that users will learn to transcend the

---

[28] Just one example of a positive step in this direction is H. Rheingold's *electricminds* Web site (http://www.minds.com), at which the older BBS concept is advanced to the level of "virtual community." The categorizations of information are clear, as is happily the distinction between a book reference, say, and "just an idea."

[29] Shepherd et al., 1996, p. 78.

blurred quality spectrum of digital information, to honor the profound relevance—and risks—of language, and above all to beware the essence of chaos in the manifest structure of the "network orange."

REFERENCES

Bajaj, S., Cheung, N., Hayward, G., and Tsai, Y. 1996. "High-speed ATM/SONET Infrastructure Research in ADTNet." *IEEE Network* 10(4), pp. 18–29.

Himmelfarb, G. 1997. "Revolution in the Library." *The Key Reporter* 62 (Spring), pp. 1–3.

Huxley, A. 1962. "Words and Their Meanings." In *The Importance of Language*, ed. M. Black. Englewood Cliffs, NJ: Prentice-Hall.

Lea, M., ed. 1992. *Contexts of Computer-Mediated Communication*. Hempstead: Harvester Wheatsheaf.

Lloyd, C. 1993. *London Sunday Times,* November 28.

Nadel, L., *et al.,* eds. 1989. *Neural Connections, Mental Computation*. Cambridge, MA: M.I.T. Press.

Oppenheimer, T. 1997. "The Computer Delusion." *Atlantic Monthly,* July, pp. 1–20.

Peters, R. 1973. "Reason and Habit: The Paradox of Moral Education." In *Philosophy and Education*. Boston: Allyn and Bacon.

Pring, R. 1971. "Curriculum Integration." *Proceedings of the Philosophy of Education Society of Great Britain* V(2), pp. 170–200.

Reid, E. 1991. "Electropolis: Communication and Community on Internet Relay Chat." Honors Thesis, Department of History, University of Melbourne, Australia.

Rheingold, H. 1993. *The Virtual Community: Homesteading on the Electronic Frontier.* Reading, MA: Addison-Wesley.

Rutter, D. 1984. *Looking and Seeing: The Role of Visual Communication in Social Interaction.* Chicester: Wiley.

Shannon, C. 1948. "A Mathematical Theory of Communication." *Bell Systems and Technology Journal* 27 (July-August).

Shepherd, D., Scott, A., Rodden T., and Vin, H. 1996. "Quality-of-Service Support for Multimedia Applications." *IEEE Multimedia,* Fall, pp. 78–82.

Tuman, M. 1992. *Word Perfect: Literacy in the Computer Age.* Pittsburgh: University of Pittsburgh Press.

Weizenbaum, J. 1976. *Computer Power and Human Reason.* San Francisco: Freeman.

VIRTUAL REALITY, AND ALL THAT

W hen a word means everything, it means nothing. Even the term *real* needs an opposite.[1]

WHAT DOES *VIRTUAL* REALLY MEAN?

This essay addresses the phenomena referred to as virtual reality (VR). We shall be more interested in the overall concept than in the specific defects or value of current VR implementations. VR is a flawed conceptual category that gives a misleading impression of what it purports to cover. The very phrase *virtual reality* is a misnomer. Much like the banner "artificial intelligence," which we addressed in Essay 2, VR as it is practiced today harbors a strong component of wishful thinking.

The word *virtual* has long had a special scientific meaning. In optical physics, for example, a virtual image is an imaginary one.[2] In other fields, *virtual* has come to mean *essentially,* or *almost,* as in the statement "Caesar

---

[1] Heim 1993.

[2] A virtual image in optics is a good example of the difference between the physically real and the physically apparent but unreal. Imagine a candle flame viewed in a mirror. One actually sees a flame appearing *behind* the mirror. Even though one could sense warmth coming from the candle image itself, this image is virtual: A hand placed *behind* the mirror—that is, right at the virtual image—will feel no heat.

was deemed a virtual god." Either way, the phrase *virtual reality* is a logical nightmare. (One would perhaps like to see VR renamed "virtuality" or some such epithet that did not clash logically with "reality.") We object to the phrase, however, because it beclouds what VR practitioners do and disguises what they don't—and can't—do.

Reality consists of the items that make up the objective order of the world. The sciences study its separate parts, establishing their occurrence and the causal relationships among them and separating fancy from reality. Whenever we say that a state of affairs *seems* to be the case or *appears* to be the case, we invoke the notion of reality by questioning whether it really *is* so. Delusions, illusions, fancies, and psychotic states are so labeled precisely because they are departures from reality. There can be disagreements about the specific content of reality, but the refusal to distinguish in general terms between reality and appearance is fatal to the successful conduct of scientific inquiry. Getting from what seems to be the case to what is the case is after all what science is about.

The implication that VR is a good approximation of reality carries a cavalier disregard for traditional notions of reality by commingling illusion, delusion, and fancy with reality. VR research can play to certain scientifically irresponsible temptations, as exemplified in this kind of observation: "It would be good to feel molecules with my hands" (Rheingold 1991). There lurks the unfortunate implication that machinery can somehow release one from the shackles of reality, with the attendant suggestion that this would be a welcome thing. It may be going too far to say so, but we see a parallel between this "alternative descriptive reality" facet of VR and the tenets of astrology. Whether one believes in astrology or not, astrology does give the impression of explaining natural events in terms of a paradigm of predestination. Just as astrology affords a simplistic explanation of complex reality, we caution aficionados of VR that their substituted reality might also be, for better or worse, a simplification. Thus VR involves a notion of special empowerment—the sense that somehow one can, via machinery, "graduate" from reality to something better. This notion is not inherent in the technology but is a consequence of its unfortunate label. Much of the technology is beneficial, useful, and legitimate.

We are not engaging in some idle semantic dispute. Words are powerful tools needing as much care as the tools in a computer lab. They make a difference in how we approach a subject and evaluate its practice and potential. Even the technologists recognize this; some of them confess a nagging dissatisfaction with the phrase *virtual reality* and propose replacing it with more felicitous language. But they do not give up the phrase, and their dissatisfaction is almost always expressed in works whose titles contain the very words. For example, Loeffler and Anderson assert at the beginning of the *The Virtual Reality Casebook* that *virtual reality* is a "buzz-word" and that a better definition would incorporate the concepts of "simulated environment" and "immersion and interactivity."[3]

Though we focus on VR, we will also have something to say about simulation and immersion. We find these notions equally questionable and believe that the line of analysis we take with VR applies to them also. The professionals we discuss in this essay are very good at what they do, but they are less good at telling others about it. In addition, part of the glamour of the field still derives from its association with reality, and if we can dissipate the glamor, we can get to a better, and more sober, appreciation of the technology itself.

VR IMPLEMENTATIONS

No agreed-upon exemplary list of VR implementations exists, but the following instances should be enough to launch our inquiry.

- An interactive map providing a tour of Aspen, Colorado. The viewer is surrounded by screens and, by touching a screen, can move from place to place, enter houses, and see how they look at different seasons of the year.
- A three-dimensional interactive version of a planned building, Sitterson Hall, at the University of North Carolina. This version became famous because its users were able to find defects in the design of a partition, establishing before the structure was built that the space would feel too cramped.
- A molecule representation through a molecular force-feedback system in which a robotic arm can manipulate the docking of molecules.
- The "Super Cockpit," of Thomas Furness, giving pilots a collection of radar, control, and weapons data, with graphical depiction of location, speed, landscape, and targets for firing. This simulation, regarded as one of the great successes of VR technology, has given rise to a number of video games.

---

[3] Loeffler and Anderson 1994, p. xiii.

- Nintendo, Atari, and Sega video games, such as Jaron Lanier's "Moondust" and the Virtuality series of Jonathan Waldern.

  Just put on the helmet and the data gloves, grab the control stick, and enter a world of computer animation. You turn your head and you see a three-dimensional, 360-degree, color landscape. The other players see you appear as an animated character. And lurking around somewhere will be the other animated warriors who will hunt you down. Aim, press the button and destroy them before they destroy you. Give it a few minutes and you'll get a feel for the game, how to move about, how to be part of a virtual world. That's virtual reality![4]

- A precursor of one type of VR in "Cinerama," wherein the wraparound screen approaches the VR dream of total immersion in the experience of a film.

- Another "total immersion" precursor of VR in "Sensorama," an early form of cinema in which touch and smell are added to sight and sound in a 3-D display.[5]

- A text-based Virtual Campus, described by the founder as follows:

  The layout of the virtual campus was modelled . . . at least, aboveground . . . on Foucault's . . . or actually Bentham's . . . panopticon, which we chose as the most effective representation we could think of for any institutional structure. The description, remember, of the layout referred to a circular building of several levels, with a central observation point. In fact, that point existed and you could quite literally stand in that room and tune into the conversations going on in any other room in the panopticon. The only difference between this panopticon and the original was the rather significant one that the student/prisoners were free to occupy that central point. And, when it came to discussing privacy rights, the panopticon provided it with some practical examples. The second plane of the campus, though, was below-ground, where we had the rhizomic steam tunnels. These tunnels were actually areas where the students were free to build anything they wanted, create their own rooms, and connect those rooms in any manner they desired to any other room in the steam tunnels. As you can imagine, the cells assigned to each student in the Panopticon very quickly developed trap doors with laundry chutes or waterslides down into the steam tunnels. Students did spend all hours of the night and day logged into the campus, talking with one another, sometimes working on class materials some-

---

[4] Heim 1993.

[5] Rheingold 1991. Rheingold describes the stereoscopic effects of this gadgetry as approaching the essential elements of VR. "I couldn't help but wonder what it would have evolved into by now if research and development had not halted three decades ago."

times not, sometimes programming useful objects for extra credit, but they spent most of that time below-ground.[6]

Cinerama and Sensorama anticipate one goal of VR, total immersion. "An *immersive experience* is one so absorbing that you cease to notice your surroundings or 'how you got there,' "[7] much like, we suggest, the experience of a serious chess player fully concentrating on the game or the effect that great works of art have on an engaged audience. The architectural environment for Sitterson Hall displays the interactive component of VR. The viewer can select which hall or room to explore and, with the help of a treadmill and handlebars, experiences the illusion of walking down corridors of the hall. "Interactivity, like immersion, is a crucial aspect of VR, the user acting to transform a computer display. There are two unique aspects of interactivity in a virtual world, navigation within the world and the dynamics of the environment."[8] Interactivity per se is also illustrated by the familiar experience of transforming the computer screen by typing on a keyboard.

All of our examples are *three-dimensional graphical displays,* except for Cinerama and the Virtual Campus. Cinerama's wrap-around screen is not three-dimensional, and the Virtual Campus is text-based and produced by an object-oriented programming language.

Computer jargon calls each of our examples simulations. "Virtual Reality is a simulated environment. The simulation may represent a real environment or it may be purely imaginary."[9] Actual facts of an actual world are simulated by the map of Aspen and the Super Cockpit, but the video games are not simulations in that sense at all. In the first sense, a good simulation incorporates the facts it supposedly simulates. If a simulation of Aspen gets the streets and the houses wrong, we would agree that it remains a simulation of Aspen, though a bad one. But even if the houses are made to fly, VR technologists call it a simulation still, and still part of virtual reality.

So far we are left with the idea that VR confines our attention to a computer-driven display, usually but not always 3-D, and simulates reality whether that reality exists beyond the screen or is purely imaginary. We

---

[6] Unsworth 1994. Unsworth had some considerable reservations about the educational effectiveness of this "virtual" setting.

[7] Pimentel and Teixeira 1995.

[8] Pimentel and Teixeira 1995.

[9] Loeffler and Anderson 1994.

might call all this virtual illusion or virtual appearance (preserving the time-honored distinction between appearance and reality). The VR use of simulation does not require verisimilitude between display and world. But there is a reason why virtual reality will continue as an inclusive label, and it is more than cultural lag. To change it would be to give up a whole raft of positive connotations. *Reality* has an impressive ring and always confers an aristocratic, quasi-scientific status. It is also reassuring and down to earth. Getting at reality just must be worthwhile. *Virtual reality* remains one of the more striking phrases invented by computer technologists, suggestive of conjuring turned to serious purpose.

THE FASCINATION WITH VR

If VR embodies a defensible concept, we should be able to discover some criteria that justify applying it to most of the items from our list of VR examples. What is needed, then, is some inquiry into the concept VR purports to express. Virtual reality fascinates people. The literature about it suggests that computers at long last have succeeded, or soon will succeed, at offering some kind of approximation to reality—an aggressive doctrine that begs an aggressive response.

The nature of VR cannot be comprehended without some agreement about the nature of the reality to which VR corresponds. Some VR proponents do say things about reality, but their remarks often sound like aimless musings about the meaning of life. Jaron Lanier, for example, asserts in an interview in *Mondo 2000* that "In Virtual Reality there's no question your reality is created by you. You made it. . . . I think being in that mode of realizing how active every moment in life is will break through the stupor."[10] Or consider this from Rheingold:

> The VR experience breaks the frame of everyday reality, although it does not, as yet, catapult the user into the kind of profoundly different experience that can be catalyzed by psychedelic chemicals. But that simple frame-breaking, and the vast potential for symbolic and what Jaron Lanier calls postsymbolic experience represented by today's crude VR systems, represents the possibility that someday, in some way, people will use cyberspace to get out of their minds as well as out of their bodies.[11]

---

[10] Cheshire 1994.

[11] Rheingold 1991.

These remarks imply that the business of VR is to make hallucinations credible by breaking "the frame of everyday reality." Other VR descriptions take a solemn tone, as though there are social ills that can be treated with substantial doses of virtual reality.

To be sure, the VR apologists are good at aphorisms and epigrams. As Chris Cheshire comments, "They compare VR with 'a work of art,' 'a dream,' 'an additional reality,' and 'a new continent.' "[12] Catch phrases like these are harmless as long as no one takes them seriously, but they do not help us grasp the reality of VR. They only substitute one group of mysteries for other mysteries.

Let us be precise about certain logical claims with respect to VR. If something is said to provide the experience of VR, the reality made virtual must itself be known and specified. Otherwise, it is an act of the imagination that, whatever its qualities, may have everything or nothing to do with reality. Of course, imagination can be a very good thing, though its use is ironic here because it so often serves to take us *away from* reality.

If we are presented with a virtual dog, we can't be sure it is a virtual dog unless we already know about the features that make for a real dog. If we have never encountered a dog or a reliable description of one, we can judge whether the barking or panting creature before us is really an outstanding achievement in virtual doghood. At the very least, the virtual dog must visually resemble a real one; better, it will bark for us, and maybe, if it is as virtually real as can be, it will lick our hand, go for evening walks, and console us when we are sad. We have the right to expect, in short, that a virtual *x* resemble in some, if not all, significant respects the real *x* to which it supposedly corresponds. Complete resemblance may be too much to expect.[13]

Thus there must be some independent identification of the real entity if we are to agree that we have its virtual version. Our assessment of the

---

[12] Cheshire 1994.

[13] The following remarks are relevant to a discussion of computer simulation. "Keep in mind that current VR systems have only a vague understanding of real physical properties. Dropping a virtual glass on a virtual floor of a kitchen doesn't cause it to shatter or make a sound unless the world designer specifically programmed it do so. Simple underlying physical principles like these are almost completely missing from current systems. . . . Simulating reality is no mean feat" (Pimentel and Teixeira 1995). But compare this: "Virtual reality is a simulated environment. The simulation may represent a real environment or it may be purely imaginary. It may follow familiar laws of physics or not; it may be filled with realistic detail or be highly abstracted" (Loeffler and Anderson 1994). Remarks like this breed confusion by allowing something that is purely imaginary to count again as a simulation.

virtual version depends on advance knowledge of the real thing being made virtual. In short, virtual reality is not self-certifying. We have to go outside the screen or the three-dimensional environment to find out whether something on that screen or in that environment has the attribute of reality.

This logical tangle is even more complicated. The application of the notion "real" is subject to certain logical conditions, one of which makes it possible for the same thing or event, depending on the description we give it, to be at once real and not real. The remarks of the distinguished Oxford philosopher, J. L. Austin are pertinent here.

> "Real" is an absolutely *normal* word, with nothing new-fangled or tech-nical or highly specialized about it. It is, that is to say, already firmly es-tablished in, and very frequently used in, the ordinary language we all use every day. Thus *in this sense* it is a word which has a fixed meaning, and so [it] can't, any more than any other word which is firmly estab-lished, be fooled around with *ad lib*. Philosophers often seem to think that they can just "assign" any meaning whatever to any word; and so no doubt, in an absolutely trivial sense, they can (like Humpty-Dumpty). . . . But whereas we can *just* say of something "This is pink," we can't *just* say of something "This is real." And it is not very difficult to say why. We can perfectly well say of something that it is pink with-out knowing, without any reference to what it is. But not so with "real." For one and the same object may be both a real *x* and not a real *y;* an object looking rather like a duck may be a real decoy duck (not just a toy) but not a real duck. When it isn't a real duck but a hallucination, it may still be a real hallucination—as opposed, for instance, to a passing quirk of a vivid imagination. That is, we must have answer to the ques-tion "A real *what*?" if the question "Real or not?" is to have a definite sense. . . .[14]

Those creatures climbing up the wall may not be real snakes, if we have had too much to drink, but they are nonetheless a real hallucination, which we do not recognize as a real hallucination. (If we did, we would not be frightened.) All of this deepens the mystery over *what form of real-ity* VR imitates, real creatures that are part of the causal order of nature, or real hallucinations that are the product of a substance-induced psychosis.

Without further clarification, there is simply no way of knowing how the purveyors of VR intend us to construe the connection between what is on the screen and what is in the world. The notion of VR is compatible with the occurrence of *anything,* because anything can, under the right de-

---

[14] Austin 1962.

scription, be a real thing of its kind. If there are real dollar bills but also real counterfeits, VR may sometimes give us real counterfeits while pretending they are real dollar bills.

FROM LITTLE REALITY TO BIG REALITY

There are of course cases where Austin's advice is not heeded, and general notions of reality are advanced. But when this occurs, there arises the same problem of the relationship of VR to reality. VR is often tied to an independent reality off the screen. Depending on the advocate, this big reality is abstract mathematics, or God, or the Absolute, or the objects of sense perception, or the substratum of the physical world. And each is labeled "reality" because it is deemed the basic stuff underlying our world and our knowledge of it.

We are left to wonder just what it is we are experiencing when the machines are turned on. Of course there are recognizable shapes, sounds, events, and scenarios. But the fact that something is recognizable does not make it real. The content of a hallucination is recognizable, but has no real counterpart. That is why it is a hallucination.

So far, our search for a fundamental meaning of VR in which reality is taken seriously has been unsuccessful. Another possible strategy is to examine what the phenomena that VR involves have in common. M. Wells, senior scientist at Logicon Technical Services, mentions, as fundamental to VR, three-dimensional objects depicted through the wearing of specially designed goggles and gloves, and the possibility of interacting with them.[15] In addition, there are the standard technical components: a display, worn like a pair of eyeglasses or as a helmet, providing the experience of viewing objects in three dimensions; a transducer, which enables the user to interact in a variety of ways with the observed environment; and an image generator. Although these are standard elements of VR technology, they say less about VR itself than about the materials of which it is constituted.

Using the technical commonalties alone to "flesh out" the VR concept would be like saying a house is a stable structure made of wood and fastened together with hammer, nails, and glue. If that were an adequate definition of a house, a book shelf, a dining room table, and a bird cage would all be houses. Rather, a house is defined with respect to the distinctive

---

[15] Wells 1991.

function it performs—as one dictionary has it, "a building that serves as living quarters for one or a few families."

A house *is* what it *does,* and it may be that VR can be evaluated in just this way—that is, with respect to a distinctive function. We shall now borrow one definition of VR, not because it is right, but because it will be illuminating to find out why it is wrong. And that will be progress in the direction of understanding what to expect of a good definition.

This is the candidate definition: "The term 'virtual reality' describes a further step in making simulated worlds accessible to the user. Virtual reality refers to the goal of allowing the user to experience computer simulations, not as columns of numbers or even graphical displays, but as environments. By wearing interface devices such as specially designed goggles and gloves, users can feel as if they are seeing and touching real three-dimensional worlds."[16] VR then presents illusions that are perceived by the user as though they were not illusions. In this definition, the same problems with the mention of "reality" arise as in our definition, but even if we waive our objections on that score, "simulation," as a distinguishing property of VR, is as problematic a conception as "reality."

SIMULATING FROM THE VACUUM

First, although proponents frequently refer to VR as a simulation medium, no limit whatsoever is placed on what is simulated. Something is called a simulation even when nothing is simulated. But not always. VR theorists also continue to use *simulation* in the standard sense of representing an independent fact or event: "As early as 1929, the Link corporation was building flight simulators. A full-sized mock-up of a fighter cockpit was constructed and mounted on a motion platform. The cockpit physically pitched, rolled, and yawed based on the pilot's actions. Results from the simple Link trainer showed that a pilot's training could be successfully simulated on an earth-bound platform."[17] Even here, simulation begins to lose its logical bearings. The pilot's training is not being simulated. The pilot is undergoing real training on an earth-bound platform that is simulating a real fighter cockpit.

Sherry Turkle preserves this replicative sense of simulation. "Just as some teachers do not want to be 'reduced' to instructing children in a

---

[16] From D. Friedman, "The Virtual Science Center" in Pickover 1992.

[17] Pimentel and Teixeira 1995.

computer 'appliance,' many resent instruction in a learning environment that often strikes them as an overblown video game. The question of simulation is passed from preschool through the college years. Why should four-year-olds manipulate virtual magnets to pick up virtual pins? Why should seven-year-olds add virtual ballast to virtual ships? Why should fifteen-year-olds pour virtual chemicals into virtual beakers? Why should eighteen-year-olds do virtual experiments in virtual physics laboratories? The answer to these questions is often: because the simulations are less expensive, because there are not enough science teachers."[18] This is VR talk that preserves the standard sense of simulation— genuine simulation.

Rheingold also talks about flight simulators before he concludes that all of VR is a simulator.

> The head-mounted displays (HMDs) and three-dimensional computer graphics, input/output devices, computer models that constitute a VR system make it possible, today, to immerse yourself in an artificial world and to reach in and reshape it. If you had to choose one old-fashioned word to describe the general category of what this new thing might be, "simulator" would be my candidate. VR technology resembles, and is partially derived from, the flight simulators that the Air Force and commercial airlines use to train pilots. In conventional flight simulators, pilots learn something about flying an airplane without leaving the ground, by practicing with a replica of airplane controls; the "windshield" of a flight simulator is a computer graphics display screen upon which changing scenery is presented according to the course the pilot steers. . . . Virtual reality is also a simulator, but instead of looking at a flat, two-dimensional screen and operating a joystick, the person who experiences VR is surrounded by a three-dimensional computer-generated representation, and is able to move around it in the virtual world and see it from different angles, to reach into it, grab it, and re-shape it.[19]

Genuine simulation is a representation of some part of the world, but the definers of VR make any VR presentation a simulation. In the former sense, there are relevant questions about the quality of a simulation: Of what is it a simulation? How well has it simulated that something? The evaluative terms drop out when a simulation can be anything. The language of the VR world lacks a clear technical sense for simulation. It borrows on the standard sense, adds it own, and moves in the space of a

---

[18] Turkle 1997.

[19] Rheingold 1991.

single sentence from one to the other, from a world of real illusion to a world of virtual reality.

Simulation, with its authoritative ring, *must* be included in the description because it implies a natural link to authentic reality. The idea of "reality" in VR requires a connection to an extrinsic reality, and such a connection is provided by the "fact" of the extrinsic reality being mirrored in VR. As a result, even an arcade game populated with imaginary dragons and monsters must be called a simulation. Recall the remark of Loeffler and Anderson: "A simulation may be purely imaginary." Even if purely imaginary, it must still be a simulation. Otherwise, no reality, and no virtual reality. In an interesting way, the idea of reality requires that of simulation. Just as the idea of "virtual" collides with "reality," the idea of "simulation" collides with "imaginary creation."

Maps, Models, and Immersion

Maps—representations of physical locations and orientations—can be thought of as paradigm examples of simulations. Some additional light can be shed on VR simulations by thinking about maps. One of the refrains that runs through typical VR rhetoric is total immersion in the all-inclusive experience. From this point of view, the goal of a VR representation is to reproduce without omission every element of the represented environment. It is a goal doomed to failure, and some reflection about mapping, as a prime case of simulation, will reveal why this is so. In the process, we shall see why the goal of simulation conflicts with that of total immersion.

A map of the state of Oregon is a simulation of some, but only some, of its features; what features are included depends on whether the purpose of the map is to give information about roads or population or topography. Of course, if the map miraculously reproduced all the features of the parent state, it would be Oregon itself, no longer a map, and still in need of one! Maps are not designed to capture all of their subject, and *the omissions are crucial to the simulation's fulfilling its purpose.* Weather maps, by the same token, are not held in contempt because they do not turn green in the spring.

M. Shubik makes the point in a slightly different way.

Given the questions, their relevance to the purpose at hand, and given an idea of the data requirements and the available time and resources, the model builder is in a position to construct his model. Possibly the basic source of all wisdom concerning the construction and use of models is that there is no such thing as an all-purpose model. The more general the set of questions posed to a scientific model, the less likely it is that the model will provide sufficiently detailed answers. An all-purpose model tends to be a no-purpose model.[20]

Consequently, the "total immersion" goal commonly assigned to VR is an exercise in futility, and the declared interest in reaching it is, for thinkers who pride themselves on their modernity, strangely reminiscent of the Romantic Age.

The immersion ideal also reflects what Cheshire calls "a cultural tenet of VR, that it is experiential. Unlike other types of computing, which require symbolic abstraction, VR allows dealing with computers to be a direct experience—therefore something familiar and attractive."[21] We shall discuss the immersion ideal later when we come to Heim and his use of Lanier. Though now is not the time to launch into an extended discussion of the role of abstractions in scientific inquiry, this much needs to be said: Immersion in the life of the senses may have a number of purposes, but one of them cannot be the systematic acquisition of knowledge. Systematic inquiry requires a certain detachment from experience, and the theories it generates are rife with concepts that have no counterpart in the sensible world, all the way from mass-point and instantaneous velocity in physics to the id and superego of Freudian psychology. Neither endless experiences nor endless summaries of experiences are the key to scientific understanding. If they were, we might all have developed Newton's laws of motion by watching apples fall from trees or Kepler's laws of planetary motion by gazing at the night sky.

Some recreational relief may come from immersion in VR-induced sensations, a kind of electronic hot tub, but immersion is not by itself an avenue to any sophisticated and cumulative understanding of the world, nor is there any verbal sleight of hand that will make it so. Thus, the physics and astronomy of so redoubtable a thinker as Aristotle suffered, contrary to common belief, from too great a dependence on experience. S. Toulmin and J. Goodfield remark that

> [Aristotle's] conclusions fitted admirably with our common experience of the world, and in many respects further experiment and observation would have reinforced his ideas rather than refuting him. Far from being

---

[20] Shubik 1981.

[21] Cheshire 1994.

a piece of armchair imagining, his theory of motion was if anything too close to the facts and not abstract enough. By contrast, when Galileo and Newton stood back and looked at the world afresh from the safe distance of a mathematician's study, they were led to re-shape dynamical theory in a form which affronted common experience.[22]

The idea that experiencing and sensing bring about understanding has an anti-intellectual quality. It demeans the cognitive role of abstract conceptions and ideas, including those that provide a theoretical underpinning for the structure of machine languages and the creation of VR itself. E. Nagel puts it this way:

> As a matter of psychological fact, elementary sense data are not the primitive materials of experience out of which all our ideas are built like houses out of initially isolated bricks. On the contrary, sense experience normally is a response to complex though unanalyzed patterns of qualities and relations; and the response usually involves the exercise of habits and of interpretations and recognition based on tacit beliefs and inferences, which cannot be warranted by any momentary experience. Accordingly, the language we normally use to describe our immediate experiences is the common language of social communication, embodying distinctions and assumptions grounded in a large and collective experience, and not a language whose meaning is supposedly fixed by reference to conceptually uninterpreted atoms of sensation.[23]

However carelessly the distinctions among illusion, fantasy, and representation are observed in VR, even VR devotees may admit to salient differences between a court document reporting the arrest of a criminal and a group of children playing "cops and robbers," or between a dynamic map of the constellations and another episode in the astral travels of Captain Kirk. If in VR we believe that every computer-generated experience somehow simulates independently existing entities or events, we begin to populate the world with goblins.

Of course, simulation is not the only potentially defining characteristic that VR specialists have accorded to their discipline, and neither is total immersion, interactivity, or any combination of these. That is because, as Michael Heim maintains, none of them, individually or together, clarify the notion of VR.

---

[22] Toulmin and Goodfield 1961.

[23] Nagel 1987.

What is virtual Reality? A simple enough question. . . . Reach for a dictionary. Webster states: Virtual: being in essence or effect not formally recognized or admitted. Reality: a real event, entity, or state of affairs. We paste the two together and read: "Virtual reality is an event or entity that is real in effect but not in fact." Not terribly enlightening. . . . The dictionary definition does, however, suggest something about VR. There is a sense in which any simulation makes something real that in fact is not. The Virtuality game combines head-tracking, glove, and computer animation to create the "effect" on our senses of "entities" moving at us that are "not in fact real." But what makes VR distinctive? "What's so special. . . ," our questioner might ask, "about these computer-animated monsters? I've seen them before on television and in film. Why call them 'virtual realities'?". . . . OK, so what is it? The next reply must be more informed: "Go to the source. Find the originators of this technology; ask them. For twenty years, scientists and engineers have been working on this thing called virtual reality. Find out exactly what they have been working to produce. When we look to the pioneers, we see virtual reality going off in several directions. The pioneers present us with at least seven divergent concepts currently guiding VR research. The different views have built camps that fervently disagree as to what constitutes virtual reality. Here is a summary of the seven: 1. Simulation—Computer graphics today have such a high degree of realism that the sharp images evoke the term virtual reality. . . . The realism of simulation applies to sound as well. Three-dimensional sound systems control every point of digital acoustic space, their precision exceeding earlier sound systems to such a degree that three-dimensional audio contributes to virtual reality. 2. Interaction—Some people consider virtual reality any electronic representation with which they can interact. . . . Defined broadly, virtual reality sometimes stretches over many aspects of electronic life. It includes the entertainer or politician who appears on television to interact on the phone with callers. It includes virtual universities where students attend classes on line, visit virtual classrooms, and socialize in a virtual cafeteria. 3. Artificiality—As long as we are casting our net so wide, why not make it cover everything artificial?. . . . But once we extend the term "virtual reality" to cover everything artificial, we lose the force of the phrase. When a word means everything it means nothing. 4. Immersion—According to this view, virtual reality means sensory immersion in a virtual environment. 5. Telepresence—Robotic presence adds another aspect to virtual reality. To be present somewhere yet present there remotely is to be there virtually (!). 6. Full Body Immersion—The interaction of computer and human takes place without covering the body. The burden of input rests with the computer, and the body's free movements become text for the computer to read. 7. Networked Communication—Because computers make networks, VR seems a natural candidate for a new communications medium. . . . Virtual worlds, then, can evoke unprecedented ways of sharing, what Jaron Lanier calls "post-symbolic communication."[24]

Lanier, on whom Heim depends heavily, is particularly lyrical in expressing his views about VR.[25] As the voice of "Networked Communication," he goes on to say that ". . . communication can go beyond verbal or body language to take on magical, alchemical properties. . . . Consciously constructed outside the grammar and syntax of language, these semaphores defy the traditional logic of verbal and visual communication." Now this kind of verbal mystification hardly seems a candidate for clarifying anything, particularly when it appears that Lanier and Heim, who takes Lanier's views with special seriousness, are already edging into the realm of "post-symbolic communication."

At this point, it will be helpful to turn to a cooling draught of prose that goes to the matter of Heim's method of consulting practitioners in a field for conceptual clarification of what they are doing. The end result of culling remarks from a Lanier to explicate an otherwise obscure concept should be patently unenlightening. C. Broad explains why this so.

> I should not expect to get much useful information [about the clarification of a practice in physics] by asking the scientist himself. If he had not a philosophical training, he would probably not understand the question or see the point of asking it, and he would certainly not have the technical equipment to answer it intelligibly. If he had a philosophical training, the chances are that he would not be a first-rate working scientist; and, even if he were, he would probably be committed to some particular (and often already exploded) philosophical view, which would bias his answers. In fact, those two modern oracles, the "plain man" and "the working scientist," resemble in one respect their ancient forerunners. The artless prattling of the former and the sophisticated technicalities of the latter stand in as much need of expert interpretation as did the inspired ravings of the Pythia at Delphi or the Sibyl at Cumae.[26]

Heim concludes that because none of the seven proposals from the pioneers yields up a statement of the nature and ultimate purpose of VR, a search must be conducted for its "esoteric essence." The essence will de-

---

[24] Heim 1993. Whatever criticisms we level at Heim should not discredit the value of his achievement in addressing the nature of VR. He takes the notion seriously, and his book-length treatment of VR is a systematic and extended effort to arrive at a clear idea about the notion. Every discussion of the subject is indebted to him.

[25] For his views on a variety of VR issues, see Lanier 1991.

[26] Broad 1961.

rive from the shared vision of all (some?) of its creators. "Perhaps the essence of VR ultimately lies not in technology but in art, perhaps art of the highest order. Rather than [to] control or escape or entertain or communicate, the ultimate promise of VR may be to transform, to redeem our awareness of reality—something that the highest art has attempted to do and something hinted at in the very label *virtual reality*, a label that has stuck, despite all objections and that sums up a century of technological innovations. VR promises not a better vacuum cleaner or a more engrossing communications medium or even a friendlier computer interface. It promises the Holy Grail."[27] As Heim has it, the esoteric essence emerges from the writings of William Gibson (in particular, his *Neuromancer*[28]), and the Holodeck of television's "Star-Trek: The Next Generation" represents "the ideal human–computer interface." Heim keeps to the idea that virtual reality must connect with an external reality of great importance. Hence the esoteric essence.

THE HOLY GRAIL

The goal of VR is to produce a version of ultimate reality through the ultimate work of art, a technological "Parsifal" that, for Heim, would be "Wagner's Holodeck." All the technical devices already available from VR will be directed to achieving this goal. For example, "VR systems," as Jaron Larnier points out, "can reduce apathy and the couch-potato syndrome simply by requiring creative decisions. Because computers make VR systems interactive, they also allow the artist to call forth greater participation from users. Whereas traditional art forms struggle with the passivity of the spectator, the VR artist finds a controlled balance between passivity and activity. The model of user navigation can be balanced by the model of pilgrimage and sacred awe."[29]

Now to comment on this extraordinary delineation of the "esoteric essence" of VR. First, such delineation constitutes a tacit admission that any coherent account of VR is somehow beyond our reach, and then it

---

[27] Heim 1993. Heim is right about one crucial matter. The label of Virtual Reality has stuck despite all objections, and the important thing is to understand why.

[28] It is instructive that Gibson, whose novel is cited often in VR literature, did not see it at all as anticipating a cyberspace-filled future but as a social critique of a present society in which technology already exerted a significant and unfortunate influence. Even so, the VR specialists who eschew literary criticism perceived themselves as fulfilling Gibson's positive vision. See Gibson 1989.

[29] Heim 1993.

compounds the problem by inventing a novel conceptual device, the "esoteric essence," that will somehow allow what sober reason forbids. This new form of verbal juggling offers advantages. It gives one complete freedom to conjure up any elements of one's own choosing and then put them into the context of this esoteric essence. After all, not much in the way of evidence can be expected to buttress the formulation of an essence as secret and elusive as this one. Ironically, the decision to embrace this expedient constitutes an admission, from a sympathetic observer of VR, that the notion of virtual reality does not stand up to critical scrutiny. Nevertheless, it is to Heim's credit that he understands the importance of clarifying the notion and struggles mightily to do so. It is further to his credit that he makes transparent the conceptual problem of dealing with virtual reality. If we take the "reality" part seriously, as does Heim, then we keep the label and hunt for holy grails. The "label has stuck, despite all objections," and Heim must find a deep fact of the world connected with the reality of virtual reality. If we *don't* take the "reality" part seriously, then we give up the all-inclusive label and distinguish between virtual reality, where genuine simulation is at work, and virtual illusion, where it isn't.

Second, the remarks about the problem "with which traditional art forms struggle," that of "the passivity of the spectator," are put forward as though there had been no art history, no literary criticism, no aesthetics, no sustained thinking about the arts until Lanier and Heim began worrying about "couch potatoes." For example, there is a sophisticated and ever-growing literature pertinent to the nature of the aesthetic spectator and the form of involvement that the arts demand of, and elicit from, an audience. The literature recognizes that the conditions of involvement may vary widely, depending on whether the audience is for poetry, drama, music, painting, or ballet.[30] To harbor the thought that VR has the capacity, in its esoteric essence, to solve the "problems" of the arts without recognizing that intelligent, perceptive thinkers have with some success been reflecting on these problems for centuries, have made distinctions that any competent student of the arts needs to address, and have proposed conclusions for which actual arguments have been given is the height of arrogance. It is as if the new crop of electronic out-thinkers were so confident of their own abilities, so pleased with the asset of ill-educated innocence, that no attention need be paid to a tradition of thought that cannot yet accommodate Holodecks.

---

[30] Any anthology in aesthetics will do. For example, see Dickie et al. 1989.

Third, the reader will recall the particular mission that the esoteric essence of VR grants to the arts. Rather than to control or escape or entertain or communicate, the ultimate goal of VR may be to transform, to redeem our awareness of reality—something that the highest art has attempted to do and something "hinted at in the very label 'virtual reality.' " Whether the "highest art" has attempted to transform our awareness of reality is open to question, unless the highest art is itself defined as art that makes such an attempt (in which case, any disagreement with this thesis is already precluded). In addition, it begins to sound as if art has been assigned the task customarily performed by religion. If this be so, it is possible to conclude that the highest art is the Old Testament and the Gospels, a position once taken by Tolstoy.[31] But this will be no problem for the VR publicists. Because they aspire to fuse all the arts into one, it may be even better if all of religion is added to the mix.

Fourth, the supporters of Heim's position on the goal of VR might be taken aback by one assessment of how well computers have performed just in the graphic arts. N. Negroponte begins a remarkable essay with this paragraph:

> Rarely have two disciplines joined forces to bring out the worst in each other as have computers and art. A mixture of mathematical exercises has predominated in the search for ways to use computers in general and computer graphics in particular for the purpose of achieving a new art form, or simply art, or both. The symmetry and periodicity of the Lissajous figures (easily generated curves on TV screens), transformations into and out of recognizable patterns, and the happenstance of stochastic processing epitomize the current palette of gadgetry used by either the playful computer scientist or the inquiring artist in the name of art. While the intentions may be good, the results are predominately bad art and petty programming. In almost all cases the signature of the machine is far more apparent than the artist's.[32]

If Negroponte is right, it may be that before reaching for the Holy Grail, VR should see what it can do with a plain water tumbler.

Apparently, we need not worry that VR will set a new standard for the arts, or achieve the goal of its esoteric essence, if the separable phenomena of VR are measured on their merits (which are considerable) and without any recourse to the rhetoric of the new visionaries. But the rhetorical visionaries of VR can be a little dangerous, because a guileless public might

---

[31] Tolstoy 1930.

[32] Negroponte 1981.

believe them and float off into the cosmic fog of virtual reality, of a "new reality," of "artificial reality," all in the name of complete immersion. The more outspoken of the VR visionaries talk as though they will change a world that they neither study nor understand. They talk of ushering in a New Age in language every bit as colorful and revolutionary as that of the Synoptic Gospels. They talk of redeeming the world with their machines yet give no evidence of knowing anything about the world outside the machine paradigm. Consider the following comment from a participant in an Internet group:

> Alan Wexelblatt writes: "We've got the world in the palm of our hand; it's a little hard not to feel like gods and goddesses." Now I'll make the most biased statement I can think of: from the extreme SE [software engineer] point of view, it looks like we're doing work which will shape the world for decades, both in personal and global terms. What are you doing that compares to that? Why should we pay attention? ". . . Because everything is more fun with an intellectual pedigree."[33]

It is hard to rein in gods and goddesses once they amass a believing parish. Once the parish has faith, it forgets that the gods and goddesses are also subject to the realities of life, including technological reality.

Some VR proponents, convinced that they understand more than they do, frequently express contempt for those who believe that values, goals, and the leavening quality of culture in a society are not necessarily technology based. Most VR proponents are not like this. Most of them write programs and sober handbooks. But they are not the dangerous ones.

The dangerous proponents tend to imply that they will succeed because the development of their technology and the universal acceptance of their goals are inevitable. Their success, as they see it, comes not only from the use of their products but also from the societal goals the product will inevitably achieve. They could be wrong about that. One way to perceive them as wrong is simply to remember how dangerous people can be if they are arrogant and messianic, and resolve as we can to resist their ardor and contain their excesses. It behooves us to be willing to put their VR products to use, when such wares indeed have a use, while guarding against the short-sighted, the narrow-minded, and the messianic whose ardor may lead to excess.

---

[33] Handelman 1991.

GVR

We close with some constructive remarks about what is and what could be good about VR. In our view, it would be epistemologically healthy to contemplate a special category, which we shall call grounded virtual reality, or GVR. Here is a good example of GVR: Though one might raise philosophical and even ethical objections to, say, "virtual surgery," in which a surgeon dons a cyber-helmet and manipulates shining, graphically rendered scalpels, one can imagine a GVR manifestation; virtual surgery would be more legitimate, justifiable, and even useful *if the patient at hand were actually represented in the supporting database.* The system would be grounded in reality in the obvious sense, and the platform might offer valuable surgical training, even per-patient preparation. The grounding notion might seem trivial, but at least with a grounded database it would be less likely that virtual surgery systems would accidentally bias themselves against certain classes of patients, such as those patients whose particular diseases are difficult to simulate. One might not have to hear, "We are sorry, but your variant of Hodgkin's disease requires too much memory."

One noble GVR manifestation is virtual-senses apparatus for the disabled. We speak not of the obvious benevolent possibilities but of the simple fact that, for a blind subject to be able to grasp the layout of a room, the system designer must be kept honest by every seeing person's notion of visual reality. The operative word is *honest.*

We can contemplate GVR scenarios that offer environmental or material advantage. There might be, for example, improved (that is, reduced) environmental impact. A good idea is the "virtual wind tunnel" concept, in which a virtual airplane wing has graphical, colored streamers indicating air flow contours.[34] This could be dubbed a mere simulation, but invite a human user to don a cyber-helmet and undergo wing control aspects of pilot training, and you have GVR. Contrast this to the noisy, electrically ravenous, and expensive classical wind tunnel. Valuable metal resources used to build many test wings might be conserved, and so on. We also like this example because it is a legitimate application of color graphics, neither for entertainment nor reality distortion, but presumably for proper scientific visualization of aerodynamic effects.

---

[34] As described in Pimentel and Teixeira 1995, p. 209.

We have recoiled already at the notion of VR force control, as in the feeling of molecules. Not so objectionable is, for example, the "virtual pottery" system of M. Sato at the Tokyo Institute of Technology.[35] We do not think of the force-feedback, virtual potter's wheel with graphical pot as true pottery in any sense. But apparently it is helpful to be able to learn pottery techniques, at least initially, without clay. The force-feedback VR, if it mimics well the hand control that a real potter enjoys, is welcome. For here, there is no implicit notion that one is feeling something that is intrinsically unfeelable.

In closing, we rejoice in the observation that a (perhaps accidental) by-product of VR research is that, finally, artisans, politicians, and writers have turned their attention to technological issues with unprecedented vigor. It is as though the most outlandish promises of VR proponents have "turned heads" and invited in this way sharper analyses of technological implications. Consider the lament of Richards and Tenhaaf:

> As artists who have worked in the area of new technologies for many years, we are confronted with limitations in the discourses around culture and technology. These are generally not formulated from the point of view of art practitioners, nor do theoretical premises active in the art world interact with technological development.[36]

These authors go on to describe an organized artist's residency for the sounding of such issues. This is a welcome and epistemologically healthy perspective, and it is the kind of perspective that the emergence of VR has instigated.

Then, too, ethical questions are being analyzed with unprecedented fervor.[37] Though it has been said that "virtual bullets cannot kill," it is becoming difficult to believe that military virtual war games are safe in the long run, what with the new war strategies that are likely to result. Given our philosophical objections to the tenets of VR, we welcome arbitrary doses of ethical, political, and artistic scrutiny. Perhaps eternal vigilance is the price we must pay to enjoy the benefits of virtual reality in its genuinely useful forms.

---

[35] As described by Hattori 1994, p. 134.

[36] Richards and Tenhaaf 1994, p. 191.

[37] A representative sampling appears in Rheingold 1994, pp. 214–217.

REFERENCES

Austin, J. 1962. *Sense and Sensibilia*. London: Oxford University Press.

Broad, C. 1961. "A Reply to My Critics." In *The Philosophy of C.D. Broad*. New York: Tudor.

Cheshire, C. 1994. "Colonizing Virtual Reality." *Cultronix* I, Number 1.

Dickie, G., Sclafani, R., and Roblin, R. 1989. *Aesthetics: A Critical Anthology*. New York: St. Martin's Press.

Gibson, W. 1989. "High Tech Life: William Gibson and Timothy Leary in Conversation." *Mondo 2000*.

Handelman, E. 1991. "sci.virtual-worlds," an Internet group.

Hattori, K. 1994. "Looking at Cyberspace from Japan." In *The Virtual Reality Casebook*, ed. C. Loeffler and T. Anderson. Princeton, NJ: Van Nostrand.

Heim, M. 1993. *The Metaphysics of Virtual Reality*. New York: Oxford University Press.

Lanier, J. 1991. In speech, from notes taken by Sonia Lins, sci.virtual-worlds, an Internet group.

Loeffler, C., and Anderson, T. 1994. *The Virtual Reality Casebook*. Princeton, NJ: Van Nostrand.

Nagel, E. 1987. *The Structure of Science*. Indianapolis: Hackett Publishing Co.

Negroponte, N. 1981. "The Return of the Sunday Painter." In *The Computer Age: A Twenty-Year View*, ed. M. Dertouzos and J. Moses. Cambridge, MA: M.I.T. Press.

Pickover, C., ed. 1992. *Visions of the Future: Art, Technology, and Computing in the Twenty-First Century*. New York: St. Martin's Press.

Pimentel, K., and Teixeira, K. 1995. *Virtual Reality: Through the New Looking Glass*. 2nd ed. New York: McGraw-Hill.

Rheingold, H. 1991. *Virtual Reality*. New York: Summit Books.

Rheingold, H. 1994. "Ethical Questions Posed by Virtual Reality Technology." In *The Virtual Reality Casebook*, ed. C. Loeffler and T. Anderson. Princeton, NJ: Van Nostrand.

Richards, C., and Tenhaaf, N. 1994. "The Bioapparatus Residency and Virtual Seminar." In *The Virtual Reality Casebook*, ed. C. Loeffler and T. Anderson. Princeton, NJ: Van Nostrand.

Shubik, M. 1981. "Computers and Modeling." In *The Computer Age: A Twenty-Year View*, ed. M. Dertouzos and J. Moses. Cambridge, MA: M.I.T. Press.

Tolstoy, L. 1930. *What Is Art?* London: Oxford University Press.

Toulmin, S., and Goodfield, J. 1961. *The Fabric of the Heavens: The Development of Astronomy and Dynamics*. New York: Harper.

Turkle, S. 1997. "Seeing Through Computers: Education in a Culture of Simulation." *The American Prospect* 31, pp. 76–82.

Unsworth, J. 1994. "Constructing the Virtual Campus." Paper delivered at the 1994 meeting of the Modern Language Association.

Wells, M. 1991. "A Report Primer on VR." <http://168.95.200.11/TechForum.dir/VR.dir/primer.html>.

EDUCATION
BE NOT AUTOMATIC

Should Teddy Roosevelt have called for a telephone on every school desk and an operator in every classroom because Alexander Bell's grand invention was changing American society? Maybe "telephonic literacy" should have been enshrined in the 1910 school curricula? Would connecting America's classrooms have radically improved the quality of our great grandparents' education and better prepared them for the rigors of the marketplace? No.[1]

EDUCATION PURSUANT TO TECHNOLOGY

In this essay, we turn to an issue of very great social importance: the effects of the computer age on education. We have touched on it elsewhere, but here it will be our center. Our first question is "What is the role, if any, of computer technology in a liberal education?" We ask this question because we reject any *a priori* belief that technological progress is the inevitable handmaiden of educational progress. Some readers may not consider this question paramount, relevant, or even interesting. But we insist it is all of these. This collection of essays came about largely because we authors felt some manifestation of Steinitz's "artistic horror" when, as

---

[1] Schrage 1997.

educators, we confronted the personal computer's saturation of college environments in the 1980s. We believe that educators everywhere, especially those with long experience in the liberal arts, have for too long been subject to incremental educational revolutions, each touted as the "next new thing" that would benefit education.

We begin our discussion of the computer's role in liberal education with the views of P. Hirst.

> The phrase "liberal education" has today become something of a slogan which takes on different meanings according to its immediate context. . . . Whatever else a liberal education is, it is not a vocational education, not an exclusively scientific education, or not a specialist education in any sense. The frequency with which the term is employed in this way certainly highlights the inadequacies of these other concepts and the need for a wider, and in the long run, more worth-while form of education. . . . Whatever vagaries there have been in the use of the term, it is the appropriate label for a positive concept, that of an education based fairly and squarely on the nature of knowledge itself, a concept central to the discussion of education at any level. . . .[2]

Paradigm examples will be like this:

> (1) They each involve certain central concepts that are peculiar in character to the form. For example, those of gravity, acceleration, hydrogen, and photosynthesis characteristic of the sciences; number, integral, and matrix in mathematics; God, sin, and predestination in religion; ought, good, and wrong in moral knowledge. (2) In a given form of knowledge these and other concepts form a network of possible relationships in which experience can be understood. As a result the form has a distinctive logical structure. For example, the terms and statements of mechanics can be meaningfully related in certain strictly limited ways only, and the same is true of historical explanations. (3) The form, by virtue of its particular terms and logic, has expression or statements (possibly answering a distinctive type of question) that in some way or another, however indirect it may be, are testable against experience. . . . A liberal education approached directly in terms of the disciplines will thus be composed of the study of at least paradigm examples of all the various forms of knowledge. This study will be sufficiently detailed and sustained to give genuine insight so that pupils come to think in those terms, using the concepts, logic, and criteria accurately in the different domains.[2]

---

[2] Hirst 1965.

What Education Is and What It Is Not

Hirst's definition can be made more concrete if we remember how all of us tend to use the notion of education. In normal usage, we distinguish it from training and specialization, in which learning also occurs. Students can learn without becoming educated. For instance, we say that so and so is a well-trained doctor but not an educated person. We do not take this to imply that the person is not a good doctor. He may be as good as can be at the business of doctoring, but he lacks something important, the same lack to which we call attention when we say that a person is a well-trained potter but not an educated person or is a well-trained laboratory technician but not an educated person.

One thing the well-trained but uneducated person might lack is sufficient understanding of the set of principles, the conceptual scheme, that undergirds the practice of doctoring or potting or lab testing—what Hirst calls "forms of knowledge." It is ironic that one can perform an activity very well, exhibit great skill at it, and still have not the slightest idea of the organizing pattern, the basic knowledge, from which comes the "theory" that its successful practice realizes. This is why well-trained television repairpersons need not be physicists and hospital laboratory technicians do not have to be theoretical biologists. It reflects the crucial distinction between education and training, between understanding a field and applying it We train acrobats, but do not have to educate them.

Consider a student who in four years of college takes nothing but physics courses. This individual emerges as a highly trained physicist, to be sure, but, for yet another reason, not an educated person. In this case, the student does not lack understanding of the deep theoretical principles. Rather, her defect is specialization. She knows little of anything else and therefore cannot grasp the connections between what is done in physics—however well understood—and anything else. That makes our student a well-trained physicist but not an educated person.

Much more remains to be said about the concept of liberal education, but we will not say it here. We want only to make the obvious point that before we decide whether something is part of education, or of liberal education, we must have some idea what education is, and that idea should be well founded. We cannot simply define as education something that we like or consider a good thing. In particular, liberal education differs from training with respect to form and breadth of knowledge. This is what distinguishes the educated person from one who is merely well trained in the practice of a skill or whose learning is confined to one domain of competence.

We hope the reader will agree that education is not the kind of catch-all notion into which we can toss anything whatsoever just because it adds this or that attractive quality. Many things can be useful or pleasing without having anything to do with education. At the beginning of this essay we posed the question of what role, if any, computer technology has to play in a liberal education. What is clear so far is that this fundamental question is not a vacuous one. There can be a wrong answer.

INCREMENTAL REVOLUTIONS

The last decades of this century are replete with talk about educational innovation. The talk is always of a new thing so very promising that it will overhaul teaching and the curriculum. Each of the new developments, from the 1960s onward, was celebrated as being at the very cutting edge of education. The litany has always run thus: after innovation, foundation grant, then adoption, then educational Nirvana. But pondering these innovations today is like touring a cemetery and taking rubbings off otherwise neglected tombstones.

The great new educational thrust of the sixties was the unstructured curriculum, suffused with relevance and unstructured because one never knew from day to day what would become relevant next. Out of this liberal imperative was to flow everything desirable: the mind of every student engaged as fully as could be with every great idea, scintillating with creativity no longer shackled by the antique manacles of curricular rigidity. This sixties motif has all but vanished. Pockets remain, but now most educators endorse a structured curriculum, no longer outmoded. Many of them are the same educators who urged the college faculties of the late sixties to embrace the new thing, to step out of the past, cast off their comfortable curricular armchairs, and dance to the music of the "new age" of education.

Another great new emergence was educational television. Television, if we could but harness its power, was to become the medium of education. No longer would instruction be constrained by the physical necessity of teacher and student inhabiting the same room. The great teachers—more important, the great minds—would be on tape, and the fundamental problem of education, how to get every student in touch with the best minds, would be solved if we could but materialize the right images. With television, the great minds and the great teachers could be everywhere. Students would no longer be condemned to such lesser versions of them as were within driving distance. And yet the television sets—the ones that still work—sit there on college campuses, used to play tapes of a Kenneth Tynan production of *Oedipus Rex* or of a monkey growing up with human parents, happily eating breakfast and throwing bananas across the room. The expected explosion of television into education turned out to be a sputter.

One more example: Not so very long ago, the latest new thing generating excitement everywhere was computer-assisted instruction. Enthusiasts promised that the essential elements of good teaching could be captured on software and that the decorative but cognitively irrelevant parts could be systematically excised. The new college, objectively tied to already established conditions of proper learning, would consist of rows of computers and rows of students, bound together by operating systems and databases. The few remaining human teachers would function as consultants, receiving any eccentric or sophisticated questions from students that the software (at least in the pioneering stages) was unable to manage.

Again, the results were not spectacular, the revolution did not arrive, and we are stuck with a very old yet apparently irreplaceable arrangement: recognizable human students and recognizable human teachers in recognizable classrooms. To be sure, some elementary and mechanical parts of some subjects lend themselves to pedagogy by program when the software is good (though often it is not): The rudiments of the vocabulary and grammar of a second language and logical Venn diagrams came to mind. But despite the heralds' trumpets, no revolution has occurred, no budgetary miracle of improving teaching by removing teachers.

Now, once more, we hear of another radical reweaving of the fabric of liberal education. But this one, we hear, is different. This one is no fad. At long last we have the genuine article: computer literacy and (in some ways more ominous) computer saturation.[3] It is now commonly believed that

---

[3] T. Oppenheimer (1997) also voices skepticism about technological innovation and education. "In 1922 Thomas Edison predicted that 'the motion picture is destined to revolutionalize our educational system and . . . in a few years it will supplant largely, if not entirely, the use of textbooks.' Twenty-three years later, in 1945, William Levinson, the director of the Cleveland public schools' radio station, claimed that 'the time may come when a portable radio receiver will be as common in the classroom as is the blackboard.' Forty years after that the noted psychologist B. F. Skinner, referring to the first day of his

this new kind of literacy is as fully a condition of being educated as is literacy itself.[4]

ENRICHING THE CURRICULUM WITH THE COMPUTER

Even the skeptical tend ultimately to surrender to the idea that the computer is central to educational effectiveness—and therefore to the corollary idea that everyone must use it because everyone's mind will be enriched by it. An educator as distinguished as Ernest Boyer echoes the prediction that "in the long run, electronic teachers may provide exchanges of information, ideas, and experiences more effectively (certainly differently) than the traditional classroom or the teacher. The promise of the new technology is to enrich the study of literature, science, mathematics, and the arts through words, pictures, and auditory messages."[5]

Boyer's hope for "exchanges of information, ideas, and experiences" in an educational context assumes, of course, the theoretical possibility of exchanges fostered by the products of software engineering. More important, such a procedure has to assume some understanding by the engineers of what devices, if made available to the technologically naive, would "enrich" the study of disciplines all the way from literature to mathematics. One must wonder how the engineers will acquire a deep and far-ranging understanding of all the disciplines they will enrich and what will happen if they do not. The hope is that engineers and teachers can collaborate and that they will be willing to do so.

If the programmers of this world are to have the effect of enriching education rather than impoverishing it, they must have or develop some sophisticated grasp of what educators expect them to be enriching. Sadly, there is not much reason to think that they either have such understanding or are inclined to develop it. We cite some remarks from computer

---

'teaching machines,' in the later 1950s and early 1960s, wrote, 'I was soon saying that, with the help of teaching machines and programmed instruction, students could learn twice as much in the same time and with the same effort as in a standard classroom.' Ten years after Skinner's recollections were published, President Bill Clinton campaigned for a 'bridge to the twenty-first century . . . where computers are as much a part of the classroom as blackboards.' "

[4] It may come as a surprise to some readers that even rudimentary literacy, let us say the widespread ability to read and write, is not ancient. Thanks to W. E. Forster's Compulsory Education Act of 1870 in England, the literacy of the British regiment ballooned, from 5 percent in 1858 to 85 percent in 1900. Thus even *that* social/political/technological revolution was relatively recent.

[5] Boyer 1984.

technologists about a conference in which they were asked to share views with a group of literary critics: "The Second International Conference on Cyberspace: Literary Criticism Collides with Software Engineering." The ostensible purpose of the conference was to generate some dialogue between literary critics and software engineers about the nature of "cyberspace." The comments of the software specialist who summarized the conference should give pause to those who look to software producers for the kind of curricular enrichment anticipated by Boyer.

> This April saw the Second International Conference on Cyberspace; it was even more colorful and controversial than its predecessors. The collected abstracts listed 98 papers, covering a wide range of topics like implementation, representation, "wiring up," AI, hermeneutics, artistry, religion, sex, fractals, cinema, anthropology, cychology [sic], ghosts, mummies, architecture. . . .

> Only 15 papers were actually presented. And, as you might expect, the content, style and state of preparation of the papers varied. Over half the presentations were given by software engineers about the cyberspaces they were building and what they learned from them. . . . The remainder of the papers were presented by academics, in the traditional language of the literary critic, examining everything from cyberspace as master narrative to a character by character analysis of Gibson's Neuromancer trilogy. I'm certain these presentations were professional enough, and I truly believe that there were some points they were trying to get across, but, frankly, I couldn't figure out what they were. After talking with other software engineers, I discovered that I was not alone.

> . . . I am one of the many software engineers in the audience who was bewildered by the language of the literary critics at this conference. Perhaps an explanation of how we think might shed some light on why. I'll use myself as an example.

> I am one of those lucky few who have actually implemented a cyberspace system and survived to tell the tale. Like many others, I have a few years of college, and lots of hands-on experience. Like many others, I don't spend much time studying the humanities or arts or reading the great French philosophers. My thought processes are instead dedicated to debugging. Debugging is usually defined as finding the failure points in a computer program, but software engineers also debug concepts and their implementations.

> . . . Time is so valuable . . . [to software engineers] . . . that several well known cyberspace implementors have stopped attending conferences— except when they can be used as advertising vehicles—in favor of getting their systems built. This trend is likely to continue if the conferences

don't offer something tangible. Presentations in the style of the literary critic aren't very tangible to us because the language used is not concrete enough for swift or accurate comprehension, extension or refutation. In short, software engineers can't debug literary criticism, so we don't get it. We can't even tell if there is any "it" to get![6]

It is disturbing to witness this kind of arrogant disdain for any discipline that does not measure up to the author's self-selected criteria for the propriety of linguistic expressions. Ironically, the author of this report, one F. Farmer, employs concepts that are fully as vague as the ones for which he berates literary criticism. J. Weizenbaum provides a brilliant explanation of this arrogance among programmers.

> It happens that programming is a relatively easy craft to learn. Almost anyone with a reasonably ordered mind can become a fairly good programmer with just a little instruction and practice. And because programming is almost immediately rewarding, that is, because a computer very quickly begins to behave somewhat in the way the programmer intends it to, programming is very seductive. . . . Moreover, it appeals most to precisely those who do not yet have sufficient maturity to tolerate long delays between an effort to achieve something, and the appearance of concrete evidence of success. Immature students are therefore easily misled into believing they have truly mastered a craft of immense power and of great importance when, in fact, they have learned only its rudiments and nothing substantive at all. . . . Just because so much of a computer-science curriculum is concerned with the craft of computation, it is perhaps easy for teachers of computer science to fall into the habit of merely training. . . . Finally, the teacher of computer science is himself subject to the enormous temptation to be arrogant because his knowledge is somehow "harder" than that of his humanist colleagues.[7]

If Farmer's remarks are characteristic and Weizenbaum's observations are right, then the future of a computer-driven curriculum is considerably less bright than Boyer envisioned. It looks instead like a curriculum whose disciplinary components will be steadily condensed into the astringent formalisms expressible in machine language and therefore ready for "debugging."[8] It should give no comfort to the humanist devotees of computer enrichment to envision their history, literature, and art cleansed of all "loose" assertions and locutions that are "not concrete enough for swift

---

[6] From a posting to the Internet newsgroup, "sci.virtual-worlds," May 10, 1991.

[7] Weizenbaum 1976, pp. 277–279.

[8] We have noticed, over years of teaching, an interesting phenomenon that involves the particular intellectual perspective of some programmers. Evidently, the constant and inex-

or accurate comprehension, extension or refutation." Of course, the price of all this would be to re-create the humanist disciplines in a form accessible to computer engineers but unrecognizable to their present practitioners. On the other hand, the humanists could help the engineers design genuinely enriching computer programs. But for this to happen, the humanists would have to understand what the programmers are doing, and the programmers would have to overcome their impatience with the language of the humanists. It can be done, and it is surely worth doing, but it will require a serious—and respectful—effort on both sides.

COMPUTER TECHNOLOGY AND LIBERAL EDUCATION

How and why has computer technology become such an integral part of a liberal education? We shall review some arguments, and we shall show that they are bad ones. Our interest, however, is not in discrediting a lot of faulty arguments but rather in suggesting something positive. Arguments to justify the inclusion of computers in liberal education cannot be like the faulty ones that follow, so we invite better justifications, if they can be developed. It is time for the cheerleading to stop and the hard analysis to begin.

Here is one argument. "Something, call it x, is bringing about a revolution in the nature of learning; therefore, x is a subject to be studied as part of a liberal education, a subject every liberally educated person needs to know." Well, the advent of the printing press incontestably revolutionized learning in Western society, and yet we feel no need to offer courses in the theory of the printing press and no need to make sure every student learns how to operate one. We feel no need for the former because printing is not a basic form of knowledge (something we have a right to expect a liberal education to bestow) and no need for the latter because being trained in the use of a printing press, however independently desirable for some, is just training and surely does not fall under the rubric of liberal education.

Neither the theory of the printing press nor training in its use is an essential or even a desirable component of a liberal education. Of course, someone must operate printing presses, but these operators need not be

---

orable effect of fatal software bugs can render a programmer acutely aware of having erred, even if in a minute way. Thus some programmers develop an acute "digital" sense of what is right and what is wrong. It is tempting to conclude that such punishment, if you will, comes to delimit any intellectual endeavor in which the programmer will later engage.

liberally educated. Of course, someone needs to make programs and run them, but from that fact alone there follows nothing whatsoever about the role of computer technology in a liberal education. Our first conclusion, simply stated, is this: Any argument for the inclusion of computer technology in a liberal education that would also justify the inclusion of printing technology in a liberal education is for that reason alone a bad argument.

Here is a second argument. "There is something, call it y, that everyone needs to know if he or she is to function and flourish in our society. Therefore, the study of y must be part of a liberal education." Well, surely the elements of good nutrition are something everyone needs to know if he or she is to function and flourish in our society, but few have argued that courses in good nutrition are essential ingredients of a liberal education or that every student must show proficiency in diet management to be certifiable as a liberal arts graduate. In one case (good nutrition), we have a putatively theoretical subject to be taught, in another case (diet management) a practice to be learned, and yet the idea of making either a required component of a liberal education seems absurd.

The fact that a subject has a theoretical component does not by itself make that subject a basic form of knowledge, even though its conclusions may be ultimately certified by such a basic form (in this case, biology). Therefore, any argument for the inclusion of computers in a liberal education cannot turn on our second argument unless we are willing to include, for reasons just as powerful, nutrition theory and diet management as essential ingredients of a liberal education.

And now a third argument. "There is something z, knowledge of which is essential to the working of the major institutions of our society. Therefore, acquiring knowledge of z constitutes an essential part of a liberal education." Accounting is such a thing, but it is not ordinarily considered part of a liberal education. A person can be trained as an accountant, but this training does not in its theoretical structure answer to the requirements of a liberal education, where the primary task is to know rather than to apply. That is why we tend to teach principles of economics in a liberal arts college, but not accounting. That is why we make a distinction between physics and engineering, between English and journalism, between biology and diet management, between psychology and career counseling: between knowledge and application. Any argument justifying the inclusion of computers in a liberal education that would also justify the inclusion of accounting in a liberal education is a bad argument.

There is even a fourth argument. "There is something, call it *b*, the use of which may from time to time be essential to someone who has mastered one of the component subjects of a liberal arts curriculum. Therefore, *b* must be included in a liberal arts curriculum." But we do not believe this either. A historian may from time have to call on epigraphy, the art of deciphering inscriptions, or paleography, the art of deciphering ancient manuscripts, without their being essential parts of a liberal education. We all know that an archeologist may have to employ radiocarbon dating in the course of his or her work. Yet we do not insist that mastery of this technique is a necessary component of a liberal education. Epigraphy, paleography, and radiocarbon dating are narrow subfields of larger fields, and such expertise can be enlisted simply by calling on a practitioner in the specialty. Any argument to justify the inclusion of computers in a liberal education that also justifies the inclusion of epigraphy or radiocarbon dating in a liberal education is a bad argument.

How Not to Teach Writing

The final argument that we will consider takes quite a different form. We paraphrase the argument like this: "Good writing is a skill whose acquisition is an essential ingredient of a liberal education. Writing can be made better, at any time, by revision. Students will be more inclined to revise, if it becomes easier for them to revise. The computer's 'cut and paste' capability makes it easier for a student to revise. A computer, in its word-processing mode, is therefore a tool crucial to the development of good writing." We mention this because it has been bandied about among teachers to the extent that it is one of the more widespread technological truisms of the 1990s.

We can do no more than point out just a few of the palpable defects of this argument. First, because of the kinds of problems that confront a student in learning how to write (those of style, analysis, argumentation, use of evidence, and so on), endless revision accomplishes little and is sometimes even debilitating. After a person has learned how to write, and thus already has a sense of what may go wrong with a paragraph or a sentence or an argument, it is convenient to be able get rid of an inferior construction and replace it with something better, and to have a word processor to do it. But emphasizing revision at the point where the skill is being developed promotes conceptual doodling more than judgment or learning.[9] Indeed, academics, who often already know how to write, and so can use iterated revision to their advantage, may conclude unreflectingly that this handy tool must have comparable advantages for the student who is learning how to write. But the things that help us when we already know how to do something are not at all the same things that help us learn how to do it.

Furthermore, if ease of revision enhances quality of writing, then writing should have taken a decisive turn for the better when the typewriter replaced the quill pen. Sad to say, we cannot convince ourselves that Gibbon would have done better with an Underwood at his side. There is no evidence to support the claim that the ease of revision that the computer affords writers has contributed in any ascertainable way to the learning of writing. The results of studies in this area are frequently inconclusive and frequently depend on questionable views of what counts as good writing.

One study, however, deserves notice, for it concludes more than that the use of a word processor is irrelevant to the quality of written prose. It suggests that certain kinds of word processors actually make discursive prose worse. M. Halio investigated *the impact of using an Apple Macintosh,* as compared to MS-DOS, on the overall quality of writing by students. She claims to have discovered this:

> Paragraphs were brief, resulting in a lack of development of thought; and sentences, too, were short, obviating the need for complex punctuation. Word choice tended to be simple, spiced with slang and colloquialisms, accenting the simplistic and generalized nature of the thought. Students were affecting a sort of pop-style of the kind found in advertising or in the mass media. . . . To test my perceptions about the childishness of the Macwriters' prose compared to the IBMers, I decided to run twenty essays randomly selected from both Mac and IBM sections through the Writers' Workbench Test Analysis program. The results confirmed my impressions: The Mac students were writing far fewer complex sentences than the IBMers. They were also using many more "to be" verbs, a sign according to composition theorists of weak and lifeless prose. Readability scores (as judged by the Kincaid scale) averaged 12.1

---

[9] Some comparable concerns about revision appear in Elbow 1981. "For one of the most frequent problems in writing, especially creative writing, is making things worse instead of better when you revise. You start out with raw writing that you know has good things in it, or perhaps you've even worked out a coherent draft and you are pleased with it strengths, its life. But obviously it needs revising. But when you finish you discover you've snuffed the life out of your piece. You've removed the problems you were trying to get rid of but somehow you've also destroyed or crippled what was good."

A Network Orange

(college level) for the IBM students, but the Mac users obtained a score of only 7.95 (slightly less than 8th grade).[10]

FROM WORDS TO PICTURES

The methodology of Halio's study has undergone vigorous criticism,[11] much of it justified. More important than whether her conclusions are valid, however, is an innovative interpretation that *welcomes* her findings as harbinger of an entirely new model of textual exposition, a form in which graphical or iconic display becomes an integral part of the written word. "Text" is redefined as a fusion of words and visual presentation, so evaluations by the old "written word" model become outmoded. An example can be found in N. Kaplan and S. Moulthrop.

> In our view, part of Halio's distress arises from a wider cultural anxiety about iconic representations. . . . As W.J.T. Mitchell argues, there is no inherent difference between depiction and description. Yet despite their inseparability and despite centuries of productive combinations, words and images have always been the scene of intense ideological struggle. . . . Not surprisingly Halio has taken up arms on the side of the verbal, at the expense of the iconic, just at the moment when the evolution of electronic technology may enable valuable combinations of words and images, to say nothing of sounds and simulations. . . . Such [electronically produced] texts require us to rethink what we mean when we say "text," suggesting that what we mean is less an object than an articulated social practice. This difference in conception would require corresponding differences in our idea of rhetoric. All this would mean rethinking the project of writing education. . . . Composition, as we envisage its future, involves more than words plus pictures, or video, or three dimensional modeling programs and the like. As Jay David Bolton argues, writing is and always has been "topographic"—it is speech made visible and then arranged in a mental space. But until graphic user interfaces for computers became available, writers could not fully exploit the spatial and visual dimensions of texts. Computer mediated technologies like hypertext allow creators of texts to construct their discourses in multiple dimensions, exploring alternative pathways for trasversal and development.[12]

---

[10] Halio 1990.

[11] For a careful and judicious critique, see Slatin et al. 1990. "Even the formalistic measures Halio uses are open to multiple interpretations, and are without sample texts for comparison; they do not by any means prove that students writing on the IBM produced significantly better work than those who used the Macintosh. Certainly these data offer no grounds for concluding that the computers caused the difference Halio perceived in the students' writing."

[12] Kaplan and Moulthrop 1990.

Here is a typical contest about computer writing where neither Halio nor Kaplan and Moulthorp are clear winners. Halio's position needs more methodological rigor. Numerical scales and intuitions about good writing are not enough. Kaplan and Moulthrop need more specificity to avoid the charge that their abstract talk about "rethinking" the notion of "text" and making it "an articulated social practice" is a convoluted way of saying that bad writing can be excused if a few pictures accompany it, qualifying the text to be judged by entirely different standards of accomplishment. Any argument that succeeds in redefining the notions of good writing or text still waits in the wings.[13]

THE SCOTTISH VERDICT: NOT PROVEN

It is time to draw a deep breath and entertain the unhappy possibility that educators may be buying arguments about the learning of writing that are without basis in fact and are born of wishful thinking. It is nearly impossible to believe that the most difficult and delicate of intellectual tasks, learning to write, could be noticeably facilitated by the availability of "cut and paste" on a computer screen or that the availability of such devices will bring on a revolution in the conception of good writing and how to judge it.

We will not continue the arguments here. There are many more, all of which lead to the same Scottish verdict, "not proven." But the considerations we have singled out and discussed require serious attention before we conclude, on the basis of argument—not fashion, not press releases, not foundation afflatus—that computerdom has a legitimate role in liberal education.

Educators need to pause and ask themselves some serious questions. What we do with computers and how we do it will depend on how well we frame our questions and how carefully we look for the truth rather than the preferred answer. If the questions are not raised, the fault lies with educators, not with the makers or programmers of computers. It is unfair to ask the makers of construction tools to design a building or to criticize them if the building is a stylistic monstrosity. The advent of computer technology challenges educators to rethink their business, an endeavor that must derive from their reflections on the nature of education itself. The purpose of this final essay has been to suggest why this is so.

---

[13] The battle continues in the journal *Computers and Composition*, beginning in 1990.

REFERENCES

Boyer, E. 1984. "Education's New Challenge." *Personal Computing,* September, pp. 81–85.

Elbow, P. 1981. *Writing with Power.* New York: Oxford University Press, p. 146.

Halio, M. 1990. "Student Writing: Can the Machine Maim the Message?" *Academic Computing,* January, pp. 16–19.

Hirst, P. 1965. "Liberal Education and the Nature of Knowledge." In *Philosophical Analysis and Education,* ed. R. Archambault. New York: Humanities Press, pp. 113–138.

Kaplan, N., and Moulthrop, S. 1990. "Other Ways of Seeing." *Computers and Composition* 7, pp. 89–102.

Oppenheimer, T. 1997. "The Computer Delusion." *Atlantic Monthly,* July, p. 1.

Schrage, M. 1997. Internet commentary: Hotflash 4.10, February 14, <hotflash-info@hotwired.com>.

Slatin, J., et al. 1990. "Computer Teachers Respond to Halio." *Computers and Composition* 7, pp. 73–79.

Weizenbaum, J. 1976. *Computer Power and Human Reason.* San Francisco: Freeman, pp. 277–279.

INDEX

A Network Orange

Toulmin S. 97–98, 107
Toy Story 7–8
transistors 6, 8
Tsai, Y. 83
Tuman, M. 83
Turing Machine 16
Turing, A. 16, 24
Turkle, S. ix, x, 94–95, 107
Tynan, K. 113

ultrabiological 21
unified multimedia 56
unitary transformation 18
University of North Carolina 87
University of Pennsylvania 5
Unsworth, J. 89, 107

Varga, R. 4, 22
Vazirani, U. 17, 21
vector quantization 27
Venn diagrams 113
Vin, H. 82–83
Virtual Campus 88–89
virtual reality x, 69, 85–87, 89, 91–92,
    95–95, 97–99, 100, 103–104,
    106–107
virtuality 86, 88

visual data 47
    display 49–50
visualization 58–61
von Neumann, J. 5, 22
von Neumann, K. 5
VR (see virtual reality)

Waldern, J. 88
war technology 3
wave function computation 18
weather prediction 29–31
Web 64, 79–81
weights 12
Weizenbaum, J. xiii, 22, 25, 39, 42, 63, 83, 116,
    123
Wells, M. 93, 107
Westberg, J. A. xvi
Wexelblatt, A. 104
Whirlwind Project 49
Wiener, N. xv, 20, 22, 40, 42
Williams, C. P. 19, 22
Woolridge, M. 34, 36, 42
World War I 3
World War II 3–5, 45, 63
World Wide Web 64–65, 70, 78, 82
Writers' Workbench Test Analysis 120
Wylde, A. xvi